Mikael Collan, Jari Hämäläinen, and Pasi Luukka (Editors)

Proceedings of the Finnish Operations Research Society 40th Anniversary Workshop – FORS40

Lappeenranta 20. – 21.8.2013

LUT Scientific and Expertise Publications No. 13, 2013

Research Reports - Tutkimusraportit
ISSN-L 2243-3376
ISSN 2243-3376

ISBN 978-952-265-435-9

ISBN 978-952-265-435-9

9 789522 654359 >

Table of Contents

Preface

This proceedings is the official publication of the Finnish Operations Research Society (FORS) 40th Anniversary Workshop. The focus of the workshop was "decision-making and optimization" and the proceedings includes twenty peer-reviewed papers that reflect the focus well.

We wish to thank all the contributors to the workshop and especially thank the workshop keynote speakers: professors Christer Carlsson, Raimo Hämäläinen, Carlos Coello Coello, Julian Scott Yeomans, Yuri Lawryshyn, and Roberto Montemanni.

In addition to the workshop a doctoral consortium is connected to the workshop, where four experts: Christer Carlsson, Julian Scott Yeomans, Yuri Lawryshyn, and Gabor Janiga share their views on some actual topics. Again thank you for your contribution.

The workshop and/or the doctoral consortium is supported by EURO, Tieteellisten Seurojen Valtuuskunta, FORS, and Lappeenrannan Teknillinen Yliopisto (LUT). The organizers thank the aforementioned for their support.

We hope that you will enjoy the workshop and wish you a pleasant stay in the city of Lappeenranta and at the Lappeenranta University of Technology.

Prof. Mikael Collan
FORS40 Local Organizing Committee Chair
FORS40 Scientific Committee Co-Chair

Prof. Jari Hämäläinen
FORS40 Scientific Committee Chair

Prof. Pasi Luukka
FORS40 Local Organizing Committee Co-Chair

A Firefly Algorithm-Driven Simulation-Optimization Approach for Waste Management Planning

Julian Scott Yeomans
OMIS Area, Schulich School of Business
York University, 4700 Keele Street
Toronto, ON, M3J 1P3 Canada
syeomans@schulich.yorku.ca

Xin-She Yang
Department of Design Engineering and Mathematics
Middlesex University, Hendon Campus
London NW4 4BT, UK
xy227@cam.ac.uk

Abstract— **Many environmental decision-making applications contain considerable elements of stochastic uncertainty. Simulation-optimization is a family of optimization techniques that incorporates stochastic uncertainties expressed as probability distributions directly into its computational procedure. Simulation-optimization techniques can be adapted to model a wide variety of problem types in which system components are stochastic. In this paper, a new simulation-optimization approach is considered that implements a modified version of the computationally efficient, biologically-inspired Firefly Algorithm. The efficacy of this stochastic Firefly Algorithm-driven simulation-optimization procedure for environmental decision-making purposes is demonstrated using a municipal solid waste management case study.**

Keywords— *Simulation-Optimization, Biologically-inspired Metaheuristic Algorithms, Firefly Algorithm*

I. INTRODUCTION

Environmental decision-making frequently involves complex problems possessing design requirements which are very difficult to incorporate into any supporting modelling formulations [1-6]. Environmental policy formulation can prove even more complicated because the various system components often also contain considerable stochastic uncertainty [7-9]. Numerous ancillary deterministic mathematical modelling methods have been introduced to support the environmental policy formulation endeavour [2,4,10]. As most environmental systems contain stochastic uncertainty, deterministic solution approaches can be rendered relatively unsuitable for most environmental policy implementations [4,7,10-14]. Consequently, environmental policy determination proves to be an extremely challenging and complicated undertaking [5,9].

Yeomans *et al*. [15] incorporated stochastic uncertainty directly into environmental planning using an approach referred to as simulation-optimization (SO). SO is a family of optimization techniques that incorporates inherent stochastic uncertainties expressed as probability distributions directly into its computational procedure [16-18]. Yeomans [8,19] has shown that SO can be considered an effective, though computationally intensive, optimization technique for environmental policy formulation.

In this paper, a new stochastic SO approach is presented that efficiently generates solutions by implementing a stochastic version of the biologically-inspired Firefly Algorithm (FA) of Yang [20-21]. For calculation and optimization purposes, Yang [21] demonstrated that the FA is considerably more computationally efficient than such commonly-used metaheuristic procedures as genetic algorithms, simulated annealing, and enhanced particle swarm optimization [22-23]. The efficacy of this approach for environmental decision-making purposes is demonstrated using the municipal solid waste management (MSW) application taken from [15].

II. FIREFLY ALGORITHM FOR FUNCTION OPTIMIZATION

While this section supplies only a relatively brief synopsis of the FA procedure, more detailed explanations can be accessed in [20-21,23]. The FA is a biologically-inspired, population-based metaheuristic. Each firefly in the population represents one potential solution to a problem and the population of fireflies should initially be distributed uniformly and randomly throughout the solution space. The solution approach employs three idealized rules. (i) The brightness of a firefly is determined by the overall landscape of the objective function. Namely, for a maximization problem, the brightness is simply considered to be proportional to the value of the objective function. (ii) The relative attractiveness between any two fireflies is directly proportional to their respective brightness. This implies that for any two flashing fireflies, the less bright firefly will always be inclined to move towards the brighter one. However, attractiveness and brightness both decrease as the relative distance between the fireflies increases. If there is no brighter firefly within its visible neighborhood, then the particular firefly will move about randomly. (iii) All fireflies within the population are considered unisex, so that any one firefly could potentially be attracted to any other firefly irrespective of their sex. Based upon these three rules, the basic operational steps of the FA can be summarized within the pseudo-code of Figure 1 [21].

Figure 1: Pseudo Code of the Firefly Algorithm

Objective Function $F(X)$, $X = (x_1, x_2, \ldots x_d)$
Generate the initial population of n fireflies, X_i, $i = 1, 2, \ldots, n$
Light intensity I_i at X_i is determined by $F(X_i)$
Define the light absorption coefficient γ
while (t < MaxGeneration)
for $i = 1$: n , all n fireflies
    for $j = 1$: n ,all n fireflies (inner loop)
        if $(I_i < I_j)$, Move firefly i towards j; **end if**
        Vary attractiveness with distance r via $e^{-\gamma r}$
    end for j
end for i
Rank the fireflies and find the current global best solution G^*
end while
Postprocess the results

III. AN FA-DRIVEN SIMULATION-OPTIMIZATION APPROACH FOR STOCHASTIC OPTIMIZATION

As described earlier, SO is a family of optimization techniques that incorporates stochastic uncertainties expressed as probability distributions directly into its computational procedure [16-18]. While SO holds considerable potential for application to a wide range of stochastic problems, it cannot be considered universally effective due to its accompanying solution time issues [16-17]. Yeomans [19] examined several approaches to accelerate the search times and solution quality of SO.

Suppose the mathematical representation of an optimization problem contains n decision variables, X_i, expressed in vector form as $X = [X_1, X_2, \ldots, X_n]$. If the objective function is represented by F and the problem's feasible region is given by D, then the related mathematical programming problem is to optimize $F(X)$ subject to $X \in D$. When stochastic conditions exist, values for the constraints and objective can often only be efficiently estimated by simulation. Thus, any solution comparison between two distinct decisions $X1$ and $X2$ necessitates the evaluation of some statistic of F modelled with $X1$ to the same statistic modelled with $X2$ [8,16]. These statistics are calculated by a simulation performed on the solutions, in which each candidate solution provides the decision variable settings in the simulation. While simulation presents the mechanism for comparing results, it does not provide the means for determining optimal solutions to problems. Hence, simulation, alone, cannot be used as a stochastic optimization procedure.

SO is a broadly defined set of solution approaches that combine simulation with some type of optimization method for stochastic optimization [16]. In SO, all unknown objective functions, constraints, and parameters are replaced by one or more discrete event simulation models in which the decision variables provide the settings under which the simulation is performed. Since all measures of system performance are stochastic, every potential solution, X, examined would necessarily need to be evaluated via simulation. As simulation is computationally intensive, an optimization component is employed to guide the solution search through the problem's feasible region using as few simulation runs as possible [8,18]. Because stochastic system problems contain many possible solutions, solution quality can be highly variable unless an extensive search has been performed throughout the problem's entire feasible domain. Population-based heuristic methods such as the FA are conducive to these extensive searches because the complete set of candidate solutions maintained in their populations permit searches to be undertaken throughout multiple sections of the feasible region, concurrently.

An FA-directed SO approach consists of two alternating computational phases; (i) an "evolutionary phase" guided by the optimization method and (ii) a simulation module. The evolutionary phase considers the entire population of solutions during each generation of the search and evolves from a current population to a subsequent one. Because of the system's stochastic components, all performance measures are statistics calculated from the responses generated in the simulation module. The quality of each solution in the population is found by having its performance criterion, F, evaluated by simulation. After simulating each candidate solution, the respective fitness values are returned to the optimization module to be utilized in the creation of the next generation of candidate solutions. In [15] a genetic algorithm (GA) was employed in optimization phase while in this paper the more computationally efficient FA is used for directing the optimization.

One primary principle of an FA is that better (i.e. fitter) solutions in the current population possess a greater likelihood for survival and progression into the subsequent generation. The new FA module evolves the system toward improved solutions in subsequent populations and ensures that the solution search does not become fixated at some local optima. After generating a new candidate solution set in the FA module, the new population is returned to the simulation module for comparative evaluation. This alternating, two-phase search process terminates when an appropriately stable system state (i.e. an optimal solution) has been attained.

IV. CASE STUDY OF MUNICIPAL SOLID WASTE MANAGEMENT PLANNING

The efficacy of this new FA-driven SO procedure will be illustrated using the municipal solid waste management planning study of Hamilton-Wentworth taken from [15]. While this section briefly outlines the case, more extensive details and the complete mathematical model can be found in [15,24]. Located at the Western-most edge of Lake Ontario, the Municipality of Hamilton-Wentworth covers an area of 1,100 square kilometers and includes six towns and cities; Hamilton, Dundas, Ancaster, Flamborough, Stoney Creek, and Glanbrook. The Municipality is considered the industrial centre of Canada, although it simultaneously incorporates diverse areas of not only heavy industrial production, but also densely populated urban space, regions of significant suburban development, and large proportions of rural/agricultural environments. Prior to the study of [15], the municipality had not been able to effectively incorporate inherent uncertainties into their planning processes and, therefore, had not performed effective systematic planning for the flow of wastes within the region. The MSW management system within the region is a very complicated process which is impacted by economic, technical, environmental, legislational and political factors.

The MSW system within Hamilton-Wentworth needed to satisfy the waste disposal requirements of its half-million residents who, collectively, produced more than 300,000 tons of waste per year, with a budget of $22 million. The region had constructed a system to manage these wastes composed of: a waste-to-energy incinerator referred to as the Solid Waste Reduction Unit (or SWARU); a 550 acre landfill site at Glanbrook; three waste transfer stations located in Dundas (DTS), in East Hamilton at Kenora (KTS), and on Hamilton Mountain (MTS); a household recycling program contracted to and operated by the Third Sector Employment Enterprises; a household/hazardous waste depot, and; a backyard composting program.

The three transfer stations have been strategically located to receive wastes from the disparate municipal (and individual) sources and to subsequently transfer them to the waste management facilities for final disposal; either to SWARU for incineration or to Glanbrook for landfilling. Wastes received at the transfer stations are compacted into large trucks prior to being hauled to the landfill site. These transfer stations provide many advantages in waste transportation and management; these include reducing traffic going to and from the landfill, providing an effective control mechanism for dumping at the landfill, offering an inspection area where wastes can be viewed and unacceptable materials removed, and contributing to a reduction of waste volume because of the compaction process. The SWARU incinerator burns up to 450 tons of waste per day and, by doing so, generates about 14 million kilowatt hours per year of electricity which can be either used within the plant itself or sold to the provincial electrical utility. SWARU also produces a residual waste ash which must subsequently be transported to the landfill for disposal.

Within this MSW system, decisions have to be made regarding whether waste materials would be recycled, landfilled or incinerated and additional determinations have to be made as to which specific facilities would process the discarded materials. Included within these decisions is a determination of which one of the multiple possible pathways that the waste would flow through in reaching the facilities. Conversely, specific pathways selected for waste material flows determine which facilities process the waste. It was possible to subdivide the various waste streams with each resulting substream sent to a different facility. Since cost differences from operating the facilities at different capacity levels produced economies of scale, decisions have to be made to determine how much waste should be sent along each flow pathway to each facility. Therefore, any single MSW policy option is composed of a combination of many decisions regarding which facilities received waste material and what quantities of waste are sent to each facility. All of these decisions are compounded by overriding system uncertainties.

Yeomans *et al.* [15] examined three likely future scenarios for the Municipality, with each scenario involving potential incinerator operations. Scenario 1 considered the existing MSW management system and corresponded to a *status quo* case. Scenario 2 examined what would occur should the incinerator operate at its upper design capacity; corresponding to a situation in which the municipality would landfill as little waste as possible. Scenario 3 permitted the incinerator to operate anywhere in its design capacity range; from being closed completely to operating up to its maximum capacity. Yeomans *et al.* [15] ran GA-directed SO for a 24-hour period to determine best solutions for each scenario representing a 10% savings from the Municipal budget. For the existing system (Scenario 1), a solution that would never cost more than $20.6 million was constructed. For Scenarios 2 and 3, Yeomans *et al.* [15] produced optimal solutions costing $22.1 million and $18.7 million, respectively.

The new FA-driven SO procedure was run on the various scenarios of the MSW problem generating the objective values shown in Table 1. Table 1 clearly indicates how the solutions produced by the new procedure are all identical with respect to their best overall cost measurements relative to the optimal solutions found in [15] for each scenario. However, this FA-driven procedure was able to generate these results in minutes as opposed to the 24-hours required in the early experimentation representing a major computational improvement.

TABLE I. BEST ANNUAL MSW PERFORMANCE COSTS (IN MILLIONS OF $) FOUND FOR (I) EXISTING SYSTEM STRUCTURE (SCENARIO 1), (II) INCINERATOR AT MAXIMUM OPERATING (SCENARIO 2), AND (III) INCINERATOR AT ANY OPERATING LEVEL (SCENARIO 3)

	Scenario 1	Scenario 2	Scenario 3
Yeomans *et al.* [15]	20.6	22.1	18.7
Best Solution Found Using FA-driven SO	20.6	22.1	18.7

V. CONCLUSIONS

Environmental decision-making problems generally possess performance specifications whose complexity are invariably compounded by stochastic uncertainties. Therefore, any ancillary modelling techniques used to support the decision formulation process must somehow be flexible enough to encapsulate the impacts from the inherent planning and stochastic uncertainties. In this paper, a stochastic FA-driven SO approach was introduced that demonstrated that (i) an FA can be effectively employed as the underlying optimization search routine for SO procedures and (ii) this approach is very computationally efficient. Since FA techniques can be adapted to solve a wide variety of "real world" problem types, the practicality of this new FA-driven stochastic SO approach can clearly be extended into numerous disparate applications and can be readily adapted to satisfy numerous other planning situations. These extensions will become the focus of future research.

REFERENCES

[1] M. Brugnach, A. Tagg, F. Keil, and W.J. De Lange, "Uncertainty matters: computer models at the science-policy interface," Water Resources Management, Vol. 21, pp. 1075-1090, 2007.

[2] A. Castelletti, S. Galelli, M. Restelli, and R. Soncini-Sessa, "Data-Driven Dynamic Emulation Modelling for the Optimal Management of Environmental Systems," Environmental Modelling & Software, Vol. 34, No.3, pp. 30-43, 2012.

[3] K.W. Hipel, and S.G.B. Walker, "Conflict Analysis in Environmental Management," Environmetrics, Vol. 22, No. 3, pp. 279-293, 2011.

[4] J.R. Lund, "Provoking More Productive Discussion of Wicked Problems," Journal of Water Resources Planning and Management, Vol. 138, No. 3, pp. 193-195, 2012.

[5] J.A.E.B. Janssen, M.S. Krol, R.M.J. Schielen, and A.Y. Hoekstra, "The effect of modelling quantified expert knowledge and uncertainty information on model based decision making," Environmental Science and Policy, Vol. 13, No. 3, pp. 229-238, 2010.

[6] W.E. Walker, P. Harremoes, J. Rotmans, J.P. Van der Sluis, M.B.A. Van Asselt, P. Janssen, and M.P. Krayer von Krauss, "Defining uncertainty – a conceptual basis for uncertainty management in model-based decision support," Integrated Assessment, Vol. 4, No. 1, pp. 5-17, 2003.

[7] J.R. Kasprzyk, P.M. Reed, and G.W. Characklis, "Many-Objective De Novo Water Supply Portfolio Planning under Deep Uncertainty," Environmental Modelling & Software, Vol. 34, pp. 87-104, 2012.

[8] J.S. Yeomans, "Applications of Simulation-Optimization Methods in Environmental Policy Planning Under Uncertainty," Journal of Environmental Informatics, Vol. 12, No. 2, pp. 174-186, 2008.

[9] H. van Delden, R. Seppelt, R. White, and A.J. Jakeman, "A Methodology for the Design and Development of Integrated Models for Policy Support," Environmental Modelling & Software, Vol. 26, No. 3, pp. 266-279, 2012.

[10] C. Fuerst, M. Volk, and F. Makeschin, "Squaring the Circle? Combining Models, Indicators, Experts and End-Users in Integrated Land-Use Management Support Tools," Environmental Management, Vol. 46, No. 6, pp. 829-833, 2010.

[11] J.M. Caicedo, and B.A. Zarate, "Reducing Epistemic Uncertainty using a Model Updating Cognitive System," Advances in Structural Engineering, Vol. 14, No. 1, pp. 55-65, 2011.

[12] L. He, G.H. Huang, and G-M. Zeng, "Identifying Optimal Regional Solid Waste Management Strategies Through an Inexact Integer Programming Model Containing Infinite Objectives and Constraints," Waste Management, Vol. 29, No. 1, pp. 21-31, 2009.

[13] B.S. McIntosh, J.C. Ascough, and M. Twery, "Environmental Decision Support Systems (EDSS) Development - Challenges and Best Practices," Environmental Modelling & Software, Vol. 26, No. 12, pp. 1389-1402, 2011.

[14] P.M. Reed, and J.R. Kasprzyk, "Water Resources Management: The Myth, the Wicked, and the Future," Journal of Water Resources Planning and Management, Vol. 135, No. 6, pp. 411-413, 2009.

[15] J.S. Yeomans, G. Huang, and R. Yoogalingam, "Combining Simulation with Evolutionary Algorithms for Optimal Planning Under Uncertainty: An Application to Municipal Solid Waste Management Planning in the Regional Municipality of Hamilton-Wentworth," Journal of Environmental Informatics, Vol. 2, No. 1, pp. 11-30, 2003.

[16] M.C. Fu, "Optimization for simulation: theory vs. practice," INFORMS Journal on Computing, Vol. 14, No. 3, pp. 192-215, 2002.

[17] P. Kelly, "Simulation Optimization is Evolving," INFORMS Journal on Computing, Vol. 14, No. 3, pp. 223-225, 2002.

[18] R. Zou, Y. Liu, J. Riverson, A. Parker, and S. Carter, "A Nonlinearity Interval Mapping Scheme for Efficient Waste Allocation Simulation-Optimization Analysis," Water Resources Research, Vol. 46, No. 8, pp. 1-14, 2010.

[19] J.S. Yeomans, "Waste Management Facility Expansion Planning using Simulation-Optimization with Grey Programming and Penalty Functions," International Journal of Environmental and Waste Management, Vol. 10, No. 2/3, pp. 269-283, 2012.

[20] X.S. Yang, "Firefly Algorithms for Multimodal Optimization," Lecture Notes in Computer Science, Vol. 5792, pp. 169-178, 2009.

[21] X.S. Yang, Nature-Inspired Metaheuristic Algorithms 2nd Ed. Frome UK: Luniver Press, 2010.

[22] L.C. Cagnina, C.A. Esquivel, and C.A. Coello, "Solving Engineering Optimization Problems with the Simple Constrained Particle Swarm Optimizer," Informatica, Vol. 32, pp. 319-326, 2008.

[23] A.H. Gandomi, X.S. Yang, and A.H. Alavi, "Mixed Variable Structural Optimization Using Firefly Algorithm," Computers and Structures, Vol. 89, No. 23-24, pp. 2325-2336, 2011.

[24] Y. Gunalay, J. S. Yeomans, and G. H. Huang, "Modelling to generate alternative policies in highly uncertain environments: An application to municipal solid waste management planning," Journal of Environmental Informatics, Vol. 19, No. 2, pp. 58-69, 2012.

An Ant Colony Approach for a 2-Stage Vehicle Routing Problem with Probabilistic Demand Increases

Nihat Engin Toklu[†], Vassilis Papapanagiotou[†], Matthias Klumpp[*], Roberto Montemanni[†]

[†]IDSIA, Dalle Molle Institute for Artificial Intelligence
USI-SUPSI, Manno-Lugano, Switzerland
Email: {engin, vassilis, roberto}@idsia.ch

[*]ild, Institute for Logistics and Service Management
FOM University of Applied Sciences
Leimkugelstrae 6, D-45141 Essen, Germany
Email: Matthias.Klumpp@fom-ild.de

Abstract—In this paper a 2-Stage Vehicle Routing Problem with Probabilistic Demand Increases is presented. We consider a simple good delivery problem where we have a retailer and many customers. The number of vehicles and the routes to use to satisfy the demands of the customers have to be optimized. The demands of the customers can probabilistically increase when the tours have started already, as it is many times the case in a realistic environment, where unexpected events happen. In such a case additional vehicles have to be sent out to cover the unserviced demand (second stage). We describe a heuristic algorithm based on the Ant Colony System paradigm and we show how different implementation of the method empirically perform on some benchmark instances. In particular, it is shown how the implementation explicitly considering the two stages of the problem produces better results.

I. INTRODUCTION

In this paper we examine the case of a 2-Stage Vehicle Routing Problem with probabilistic demand increases. In this problem we consider the case of a company that needs to serve multiple customers with demands and has to decide the routes and number of vehicles needed to serve the customers. The objective is to minimize the total distance traveled.

The number of vehicles and the routes that the companies decide to use to satisfy their customers have economic and environmental impacts, which are directly related to the distance covered by the vehicles, and the number of vehicles used itself.

Additionally, in this version of the problem the company accounts for probabilistic increases in the demands of the customers. The situation of probabilistic increases in the demand arises in many real situations one of which is caused due to the "bullwhip effect".

The supply chain adverse effect known as "bullwhip effect" has been documented in research many times for more than fifty years [1], [2]. The "bullwhip effect" or the "Forrester effect" refers to the fact that even small increases in demand can cause an all the more bigger and abrupt change as one looks further back in the supply chain.

The problem of servicing a number of customers with a fleet of vehicles is known as the Vehicle Routing Problem (VRP). In VRP usually there is a central depot of goods that must be delivered to customers that have placed orders for such goods. The objective of VRP is to minimize the cost of distributing the goods.

A related version of VRP is the Capacitated Vehicle Routing Problem (CVRP). In the classical version of the problem of CVRP, we assume that there is a depot of a product, we have customers with certain demands and a limited number of vehicles each with a limited carrying capacity to satisfy the demands of the customers. The goal is to route the vehicles to the customers with minimal travel costs. We assume that the capacity of one (single) vehicle is always greater than the demand of a single customer.

An even more related problem to the one presented in this paper is the Capacitated Vehicle Routing Problem with Stochastic Demands (VRPSD). In VRPSD where the demands are stochastic, the actual demand of the customer is revealed when the vehicle arrives at the customer location. Before the vehicle leaves the depot, its route has been predefined according to a feasible permutation of the customers, called *a priori* tour. In the classical VRPSD the vehicle, while following the a priori tour, has to choose whether to proceed to the next customer or return to the depot for restocking.

In this paper we propose a variant of VRPSD where demands can be increased with some probability. Additionally, the vehicles are filled to their capacity and always proceed to serve the customers without deviating from their planned routes. If a vehicle gets empty without having served all the demands of the customers, then a new vehicle begins from the depot to serve the remaining demands of the customers, by following a newly calculated tour. This procedure can be repeated until all the demands of the customers are met.

The rest of this paper is organized as follows: Section II

lists the works reported in the literature which are related to our study. Section III provides a detailed definition of the studied problem. In section IV, our proposed metaheuristic algorithm for solving the problem is explained. Section V lists the experimental results obtained by our algorithm. Finally, the conclusions are drawn in section VI.

II. RELATED WORK

The Vehicle Routing Problem with Stochastic Demands has been researched before in a number of papers but to the extend of our knowledge with different assumptions to our variant. Whenever not mentioned below, it is assumed that the VRPSD solved is the classical one.

In [3], the difference in assumptions is that the vehicle returns periodically to the depot to empty its current load. Uncertainty is handled by building an a priori sequence among all customers of minimal expected length. Also simple heuristics are proposed and theoretical investigations are performed. In [4] simulated annealing is used to solve VRPSD. In [5] an exact algorithm for the VRPSD is proposed by formulating it as a two stage problem and solving the second stochastic integer program using an integer L-shaped method. In [6] different hybrid metaheuristics are analyzed in terms of performance and compared to state of the art algorithms. This paper also uses the notion of 'preventive restocking'. Preventive restocking means that the vehicle chooses to go to the depot for restocking even though it is not empty and can satisfy the next customer. In [7] an Ant Colony Optimization is proposed called "neighborhood-search embedded Adaptive Ant Algorithm (ns-AAA)" in order to solve VRPSD. The VRPSD solved in this paper has the assumptions of the classical VRPSD and also uses preventive restocking. In [8] a multiobjective version of VRPSD is solved by means of evolutionary methods. The algorithm finds tradeoff solutions of complete routing schedules with minimum travel distance, driver remuneration and number of vehicles, with constraints such as time windows and vehicle capacity. In [9] duration constraints are imposed on the expected delivery costs and this affects the structure of the set of a priori tours.

III. PROBLEM DEFINITION

Let $G = (L, A)$ be a graph with L being a set of locations and A a set of arcs connecting the L vertices. We assume that the graph is complete and in the set of locations 0 is always the depot. We also let V be the set of vehicles available for delivery and c_{ij} is the cost of traveling from location i to location j.

In order to deal with probabilistic increases in the demands, we generate various scenarios and in each of them the demands are perturbed in various ways. We let S be the set of scenarios considered, that are sampled according to a Monte Carlo technique, and d_i^s the demand in location i in scenario $s \in S$.

Our objective is to find a feasible solution x that minimizes the total travel distance and therefore also the total carbon dioxide emission. We let x^v be the route decided for vehicle $v \in V$ in x, $|x^v|$ the length of the route x^v, and x_k^v the k-th visited place of the vehicle v in x.

In case the initial vehicle runs out of capacity, a new one begins from the depot to satisfy the remaining demands of the clients. The new vehicle calculates a new route using nearest neighborhood heuristic (NNH) (see [10]).

To evaluate the final cost we first evaluate the cost ignoring uncertainty which we call base cost. Then we simulate the route for all scenarios taking into account uncertainty and return the average of the cost computed. We call this the fix function and it is computed in the objective function by the term $\sum_{s \in S}(\frac{F_s(x)}{|S|})$. $F_s(x)$ is the additional cost due to stochasticity in scenario s.

To sum up, our problem can be expressed as:

$$\min \left[\sum_{v \in V} \sum_{k=1}^{|x^v|-1} (c_{ij} | i = x_k^v, j = x_{k+1}^v) + \sum_{s \in S} (\frac{F_s(x)}{|S|}) \right] \quad (1)$$

subject to

$$x_1^v = x_{|x^v|}^v = 0 \qquad \forall v \in V$$
$$x_k^v \neq x_{x'}^{v'} \qquad \forall v, v' \in V$$
$$\forall k, k' \in \{2, ..., |x^v| - 1\}$$
$$k \neq k' \text{ if } v = v'$$
$$x_k^v \in (L \backslash \{0\}) \qquad \forall v \in V, \forall k \in \{2, ..., |x^v| - 1\}$$
$$\sum_{k \in \{2, ..., |x^v| - 1\}} \underline{d}_{x_k^v} \leq Q \qquad \forall V \in V$$

The fix function $F_s(x)$ can be explained as simulating the solution x over scenario s, getting a list of unsatisfied customers, heuristically finding tours for the extra vehicle(s) by using NNH to revisit the unsatisfied customers, and finally returning the total cost of these tours. The algorithmic explanation is as follows:

function $F_s(x)$ **is:**
$unsatisfied\_customers \leftarrow []$
for all $v \in V$ **do**
   $serving\_capacity \leftarrow vehicle\_capacity$
   for all $k \in \{2, ..., |x^v| - 1\}$ **do**
      if $serving\_capacity \geq d_i^s$ **then**
         $serving\_capacity \leftarrow serving\_capacity - d_i^s$
      else
         add k to $unsatisfied\_customers$
         $missing_k \leftarrow d_i^s - serving\_capacity$
         $serving\_capacity \leftarrow 0$
      end if
   end for
end for
$SubGraph \leftarrow \{0\} \cup unsatisfied\_customers$
$SubVRP \leftarrow$ VRP problem on $SubGraph$ where
     the demands are given by $missing_k \forall k \in SubGraph$
$y \leftarrow$ solve $SubVRP$ via nearest neighbour heuristic
return travel cost of y

IV. ANT COLONY SYSTEM

The Ant Colony Optimization algorithm (ACO) first proposed in [11] and [12], is a probabilistic technique for finding solutions to difficult combinatorial optimization problems which can be reduced to finding good paths through graphs like the traveling salesman problem (TSP) or the Vehicle Routing Problem (VRP).

The basic idea of applying ACO to the proposed problem of this paper is as follows. We have a number of artificial ants that "walk" on the graph $G = (L, A)$ and each one constructs a solution. The initial ants wander randomly. When they have constructed a solution artificial pheromones trails are laid down that indicate the arcs they have walked on. The amount of artificial pheromones depends on the quality of the generated solution. Over time, pheromone trails start to evaporate. A shorter path suffers less from evaporation because it gets marched over more frequently and the amount of artificial pheromone on it becomes higher. The mechanism of pheromone evaporation helps avoiding early convergence to local optima and if it did not exist, the paths chosen by the first ants would be excessively attractive.

Subsequent ants walking on the graph to find new solutions when they encounter a pheromone trail, they are more likely to follow the arcs with higher amounts of pheromone left. There is, however, a smaller probability that the ant will follow other arcs too. We repeat this procedure iteratively and due to the pheromone amount increase in the arcs belonging to good solutions, it can be observed that the ants converge to a near-optimal heuristic solution.

In this study, we use the *ant colony system* approach, an *elitist* ant colony optimization variation. In this context, elitism means that only ants that improve the currently known best solution are allowed to lay down new pheromones. The main consequence of elitism is that ants mainly apply local searches around the best solutions.

In this paper, for the implementation, the ant colony system described in [13] is used. Firstly, the algorithm is initialized by using NNH. The best solution of NNH is stored in a variable called *best*. Then, a new ant colony is generated with Ω ants. The cost of the new solution is evaluated by the objective function found in section 1. If it is better than the one in the variable *best*, then our new solution is stored to the variable *best*. Then iteratively we repeat the procedure of generating a new ant colony and constructing new solutions until a finishing criterion is met.

In order to generate new solutions, an individual ant starts by picking locations to visit for the first vehicle, step by step. We consider that the first location picked is always location 0 which is the depot. Next, it picks a new location from the set of visitable locations. A customer locations is visitable if it is not already visited, it is reachable from the current location and if adding it does not violate the capacity constraint. We assume the ant is now on the location i and wants to go to a visitable location j. In order to pick j, the ant is influenced by the pheromone left on the arc (i, j) expressed by $\tau_{i,j}$. To quantify the influence more definitions need to be made. We consider N_w to be the set of visitable locations for ant w, d_{ij} the distance between locations i, j, $\eta_{ij} = \frac{1}{d_{ij}}$ indicating the proximity of the two locations; β is a parameter which determines the importance of proximity and finally $\alpha \in [0; 1]$ determines the probability for an ant to decide between exploitation and exploration. Thus, an ant located in i with probability α chooses to do exploitation and picks its next location j by following the arc (i, j) which maximizes $\tau_{ij} \cdot [\eta_{ij}]^\beta$. Otherwise, with probability $1 - \alpha$, it decides to do exploration. In this case, the probability of visiting location

TABLE I. THE RESULTS OBTAINED BY USING DIFFERENT APPROACHES

Instance	Approach	Cost		Vehicles	
		Average	StDev	Average	StDev
tai100a.dat	1-Stage Best-Case	2610.03	58.63	12.02	0.0
	1-Stage Average-Case	2267.09	30.64	13.44	0.04
	1-Stage Worst-Case	2371.36	39.19	14.04	0.01
	2-Stage Average-Case	2487.26	110.07	12.03	0.01
tai100b.dat	1-Stage Best-Case	2413.99	40.63	12.0	0.0
	1-Stage Average-Case	2183.65	39.38	12.46	0.05
	1-Stage Worst-Case	2175.77	33.13	13.04	0.02
	2-Stage Average-Case	2303.85	55.46	12.0	0.0
tai100c.dat	1-Stage Best-Case	1782.81	73.53	12.0	0.0
	1-Stage Average-Case	1546.38	36.42	12.47	0.05
	1-Stage Worst-Case	1618.03	21.02	13.03	0.01
	2-Stage Average-Case	1580.28	45.64	12.0	0.0
tai100d.dat	1-Stage Best-Case	1954.59	87.44	12.0	0.0
	1-Stage Average-Case	1771.12	38.6	12.49	0.02
	1-Stage Worst-Case	1747.73	20.1	13.02	0.01
	2-Stage Average-Case	1806.49	29.56	12.0	0.0
tai150a.dat	1-Stage Best-Case	3793.40	55.12	16.23	0.04
	1-Stage Average-Case	3613.45	46.85	17.61	0.06
	1-Stage Worst-Case	3693.16	47.81	18.09	0.01
	2-Stage Average-Case	3703.80	61.79	16.18	0.08
tai150b.dat	1-Stage Best-Case	3545.05	72.52	15.16	0.13
	1-Stage Average-Case	3228.53	90.93	16.62	0.01
	1-Stage Worst-Case	3356.87	106.8	17.07	0.01
	2-Stage Average-Case	3495.09	101.41	15.11	0.07
tai150c.dat	1-Stage Best-Case	3391.10	51.72	15.69	0.02
	1-Stage Average-Case	2928.72	98.75	16.72	0.03
	1-Stage Worst-Case	2946.34	94.12	17.29	0.65
	2-Stage Average-Case	3175.00	197.04	15.68	0.01
tai150d.dat	1-Stage Best-Case	3359.90	107.83	15.14	0.05
	1-Stage Average-Case	3064.09	95.21	16.62	0.07
	1-Stage Worst-Case	3061.22	33.01	17.08	0.01
	2-Stage Average-Case	3207.96	62.18	15.12	0.03

j is: $p_{ij}^w = (\tau_{ij} \cdot [\eta_{ij}]^\beta)/(\sum_{j' \in N_w} \tau_{ij'} \cdot [\eta_{ij'}]^\beta)$ if $j \in N_w$; 0 otherwise. In the present work parameters of the algorithm are set to the values suggested in [13]. For a more detailed description of the algorithm, the reader is addressed to [13].

Multiple approaches can be derived from the basic ant colony system to solve our problem, by implementing various objective functions which approximate our main objective (1). These approaches are as follows:
— 1-Stage Best-Case Approach: This approach ignores the probabilistic increases in the demands and solves the problem only by considering the base cost in the objective function of the ant colony system. In other words, it assumes that the customer demands are not increased, and then it treats the problem as a regular CVRP.
— 1-Stage Average-Case Approach: Like the previous approach, this approach treats the problem as a regular CVRP, except that each customer demand is assumed to be equal to its lowest possible value plus the expected value of the probability distribution of its related demand increase.
— 1-Stage Worst-Case Approach: This approach also treats the problem as a regular CVRP. Each customer demand is assumed to be equal to its lowest possible value plus 2 times the standard deviation of the probability distribution of its related demand increase.
— 2-Stage Average-Case Approach: This approach treats the problem as a 2-Stage problem, where in the first stage the original demand is considered. For the second stage a single scenario, where each demand increase is assumed to be equal to the expected value of its related probability distribution, is considered. In this way the behavior of a real planner, facing the realization of uncertainty, is simulated.

V. Experimental Results

In this section, we present our experimental results. For testing our approach, we used the Capacitated Vehicle Routing Problem instances provided in [14], namely, tai100{a,b,c,d} and tai150{a,b,c,d}, which are instances considering 100 customers and 150 customers, respectively. Note that, these instances were originally created for the classical CVRP where the demands are deterministic. Since, in this study, we consider that there are probabilistic increases in the demands, we modified these instances. In more details, within each considered problem instance, for each customer location $i \in (L \setminus \{0\})$, the following modifications were applied:
− accept the demand at location i as the base demand $\underline{d}_i$
− generate a random number r within $[0; \underline{d}_i \cdot R]$ where R is a parameter that we set as 0.2 in our studies
− declare that the demand at location i is stochastic according to the half-gaussian distribution, with the base demand $\underline{d}_i$ and standard deviation r. Each ant colony system implementation was implemented in C and executed on a computer with Intel Core 2 Duo P9600 @ 2.66GHz processor with 4GB RAM.

Each ant colony system implementation was executed on the considered modified instances. Over 9 runs (each run taking 3 minutes), the average costs and the standard deviations ("StDev") are reported in table I. In the table, the costs are calculated according to the objective function (1) with 10000 random scenarios sampled in a Monte Carlo fashion. This should provide a good approximation of the expected real situation. In the results, it can be seen that, in general, the 2-Stage Average-Case Approach finds solutions with shorter travel times and with minimized number of vehicles. Solutions with even shorter travel times can be provided by the 1-Stage Average-Case Approach, but these solutions require more vehicles, and this usually has a very bad impact on the total cost of the solution for the company. When we analyze the solutions of the other approaches, we see that: the 1-Stage Best-Case approach tends to provide expensive solutions, and is clearly dominated (ignoring uncertainty is not convenient); the 1-Stage Worst-Case approach provides over-protective solutions where too many vehicles are required. In conclusion, the results show that explicitly taking into account the two stages of the optimization inside the algorithm is convenient and leads to more desirable solutions, even when a very simple, polynomial-time heuristic is used for the optimization of the second stage (as in our case).

VI. Conclusions

A 2-Stage vehicle routing problem with probabilistic demand increases was discussed, where in the first stage the vehicles head out from the depot to satisfy the base demands of the customers and in the second stage extra vehicles are sent again to the customers who were not completely satisfied because of the dynamic increases in the demands. Various ant colony optimization approaches were proposed to solve the problem heuristically. From the experiments it was seen that — on the problems considered — adopting a 2-Stages algorithm that more closely sticks to the reality, produces more attractive solutions from an economical point of view.

Acknowledgment

N.E. T. was funded by the Swiss National Science Foundation through project 200020-140315/1: "Robust Optimization II: applications and theory", and by Hasler through project 13002: "Matheuristic approaches for robust optimization". V. P. was funded by the Swiss National Science Foundation through project 200020-134675/1: "New sampling-based metaheuristics for Stochastic Vehicle Routing Problems II".

References

[1] J. Forrester, "Industrial dynamics," 1961.

[2] H. L. Lee, V. Padmanabhan, and S. Whang, "Information distortion in a supply chain: the bullwhip effect," *Management science*, vol. 43, no. 4, pp. 546–558, 1997.

[3] D. J. Bertsimas, "A vehicle routing problem with stochastic demand," *Operations Research*, vol. 40, no. 3, pp. 574–585, 1992.

[4] D. Teodorović and G. Pavković, "A simulated annealing technique approach to the vehicle routing problem in the case of stochastic demand," *Transportation Planning and Technology*, vol. 16, no. 4, pp. 261–273, 1992.

[5] M. Gendreau, G. Laporte, and R. Séguin, "An exact algorithm for the vehicle routing problem with stochastic demands and customers," *Transportation Science*, vol. 29, no. 2, pp. 143–155, 1995.

[6] L. Bianchi, M. Birattari, M. Chiarandini, M. Manfrin, M. Mastrolilli, L. Paquete, O. Rossi-Doria, and T. Schiavinotto, "Hybrid metaheuristics for the vehicle routing problem with stochastic demands," *Journal of Mathematical Modelling and Algorithms*, vol. 5, no. 1, pp. 91–110, 2006.

[7] M. Tripathi, G. Kuriger *et al.*, "An ant based simulation optimization for vehicle routing problem with stochastic demands," in *Simulation Conference (WSC), Proceedings of the 2009 Winter*. IEEE, 2009, pp. 2476–2487.

[8] K. Tan, C. Cheong, and C. Goh, "Solving multiobjective vehicle routing problem with stochastic demand via evolutionary computation," *European Journal of Operational Research*, vol. 177, no. 2, pp. 813 – 839, 2007. [Online]. Available: http://www.sciencedirect.com/science/article/pii/S0377221706000208

[9] A. L. Erera, J. C. Morales, and M. Savelsbergh, "The vehicle routing problem with stochastic demand and duration constraints," *Transportation Science*, vol. 44, no. 4, pp. 474–492, 2010.

[10] D. S. Johnson and L. A. McGeoch, "The traveling salesman problem: A case study in local optimization," *Local Search in Combinatorial Optimization*, pp. 215–310, 1997.

[11] M. Dorigo, "Optimization, learning and natural algorithms (in Italian)," Ph.D. dissertation, Dipartimento di Elettronica, Politecnico di Milano, Milan, Italy, 1992.

[12] M. Dorigo, V. Maniezzo, and A. Colorni, "Positive feedback as a search strategy," Tech. Rep., 1991.

[13] L. M. Gambardella, É. Taillard, and G. Agazzi, *New Ideas in Optimization*. McGraw-Hill, 1999, ch. "MACS-VRPTW: A Multiple Ant Colony System for Vehicle Routing Problems with Time Windows", pp. 63–76

[14] E. Taillard, "Vehicle routing instances," [Online]. Available: http://mistic.heig-vd.ch/taillard/problemes.dir/vrp.dir/vrp.html, Accessed June 2013.

Solution Space Visualization as a Tool for Vehicle Routing Algorithm Development

Jussi Rasku*, Tommi Kärkkäinen* and Pekka Hotokka*
*Department of Mathematical Information Technology,
University of Jyväskylä, P.O. Box 35, FI-40014, Finland
Email: {jussi.rasku, pekka.hotokka, tommi.karkkainen}@jyu.fi

Abstract—Understanding the exploration of search space of vehicle routing problems is important in designing efficient algorithms. These spaces are multidimensional and hard to visualize, which makes it difficult to examine the trajectories of the search. We use a technique based on multidimensional scaling to visualize objective function surfaces for such problems. The technique is used to examine a full objective function surface of small VRP and TSP problem instances, and the neighbourhood as observed by heuristic search algorithms.

I. INTRODUCTION

Combinatorial optimization problems are usually inherently multidimensional. For example, the well known travelling salesman problem (TSP) has n^2 binary decision variables, where n is the number of cities to visit. Understanding, let alone visualising such solution spaces is challenging. A visualization technique that could plot the trajectory of an algorithm in search space would be a useful tool for algorithm developers (Halim and Lau, 2007).

In this study we present a visualisation technique based on multidimensional scaling (MDS). We address the limitations and challenges of the presented technique, such as the definition of the distance metric between solutions. We will also show how to use the presented visualization technique to depict two well known routing problems, the TSP and the capacitated vehicle routing problem (CVRP). To our knowledge, this is the first time this kind of technique is used to make observations of routing problem solution spaces.

In Section II we give a brief overview to the vehicle routing problem. In Section III we review previous relevant work on understanding VRP solution spaces. Section IV describes our visualization technique and Section V gives examples of the different landscapes and addresses the accuracy of the method. We conclude our study in section VI.

II. THE VEHICLE ROUTING PROBLEM

VRP is one of the most studied $\mathcal{NP}$-hard combinatorial optimization problems (Toth and Vigo, 2002). The task in VRP is to find optimal *routes* for *vehicles* leaving from a *depot* to serve a specified number of *requests*. In its archetypical form each customer must be visited once by exactly one vehicle. Each vehicle must leave from the depot and return there after serving customers on its route. Typical objective is to minimize the total length of the routes.

Following Toth and Vigo (2002) a graph formulation for VRP can be given. Let $V = 0, \ldots, n$ be the set of vertices where the depot has the index 0 and where the indices $1, \ldots, n$ correspond to the customers. The graph $G = (V, E)$ is complete with each edge $e = (i, j) \in E$ having an associated non-negative c_{ij} that is the cost of traversal from vertex i to j. Each of the edges $(i, j) \in E$ has an adjoining binary decision variable x_{ij} to decide whether traverse the edge. Using the notation of the graph formulation, a linear programming model for VRP can be given (Toth and Vigo, 2002).

$$\min \sum_{i \in V} \sum_{j \in V} c_{ij} x_{ij} \tag{1}$$

$$\text{s.t.} \sum_{i \in V} x_{ij} = 1, \quad \sum_{i \in V} x_{ji} = 1 \qquad \forall j \in V \setminus \{0\} \tag{2}$$

$$\sum_{i \in V} x_{i0} = K, \quad \sum_{j \in V} x_{0j} = K \tag{3}$$

$$\sum_{i \notin S} \sum_{j \in S} x_{ij} \geq \gamma(S) \qquad \forall S \subseteq V \setminus \{0\}, S \neq \emptyset \tag{4}$$

$$x_{ij} \in \{0, 1\} \qquad \forall i, j \in V \tag{5}$$

Different routing problem variants can derived by defining specialized objectives and by adding additional constraints. Out of these, the most usual one, CVRP is used as an illustrative example in this work. In CVRP a set of identical vehicles, with a *capacity* of C, serve requests with a non-negative demand of d_i. The problem is to find the lowest cost solution with minimum number of tours K^* so that the capacity C is not exceeded by any of the tours.

III. ON SOLUTION SPACE ANALYSIS

Statistical analysis of fitness landscapes has proven to be useful approach in understanding solution space structure (Weinberger, 1990). Previous approaches for examining routing problem search spaces, and observations of thereof, are explored in e.g. (Fonlupt et al., 1999; Kubiak, 2007; Czech, 2008; Pitzer et al., 2012). These methods involve probing the solution space in order to calculate statistical measures that describe the behaviour of the objective function.

These studies have revealed that routing problem search space is rugged, multimodal and tends to have a "big valley" structure where the local optima are clustered close to each other. Landscape ruggedness describes the expected amount of local variance of the objective function values around any given neighbourhood.

Halim and Lau (2007) and Mascia and Brunato (2010) have previously introduced tools that can be used in stochastic local search (LS) landscape visualization and analysis. The aim of these studies has been to further the understanding of the problems. Because of the differences in underlying nature of combinatorial optimization problems, we argue that problem structure exploration tools need to be adapted to different domains.

IV. DESCRIPTION OF THE USED METHOD

Biggest challenge in visualizing the search space of VRP comes from the multidimensional nature of the problem. A combinatorial optimization problem, such as VRP, may have hundreds or thousands of binary decision variables. A technique called multidimensional scaling (MDS) can be used to bring the dimensionality down to two or three dimensions. (Brunato and Battiti, 2010) Following tasks are expected:

(A) Find a solution distance (or dissimilarity) measure.
(B) Implement an effective method of enumerating the solutions for a full search space visualization.
(C) Select the MDS method.
(D) Draw the visualization.

Fig. 1: Visualization process

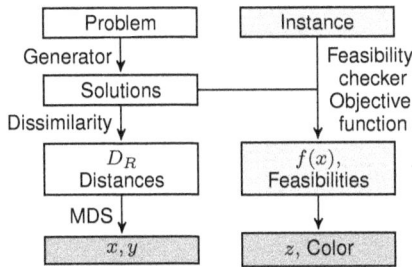

Our visualization technique uses a process that is described in Figure 1. The problem with the size n; and instance with capacity constraint, Q distance matrix D, and demands d_i; can be treated separately. For the problem, all solutions are generated and their distances to each other calculated using a dissimilarity measure. MDS produces the x and y coordinates for the visualization. Then, feasibility and objective function values are calculated for each solution of the instance. From objective values we get the z coordinate and from fitness the colour that is used to differentiate feasible and infeasible solutions.

A. Solution similarity

Let $s_1, s_2 \in R$, be two solutions of a symmetric VRP. The distance between solutions $d(s_1, s_2)$ can be defined as the number of edges *not shared* by the solutions. Thus, d is a dissimilarity measure of VRP solutions. For operators like 2-opt and relocate (Fig. 2), this Manhattan distance has been shown to be a good metric to be used with routing problems (Fonlupt et al., 1999). However, other binary dissimilarity measures such as Dice, Jaccard, Yule, Russel-Rao, Rogers-Tanimoto and Sokal-Sneath (Choi et al., 2010) can also be used, and we will present the first comparison of these measures in the domain of routing problems.

B. Enumerating VRP solutions

We needed an effective way of enumerating all solutions without capacity, time window, or similar constraints. Note that the node degree (2), (3), subtour elimination (4), and variable type (5) constraints from Section (II) are still respected. If we let $\forall K : K \in \{1 \dots n\}$, all possible solutions are created. We consider only the symmetric case where the graph is undirected. Therefore, routing problems where the direction of the tour is important (PDP, VRPTW etc.), are not considered in this study.

Fig. 2: Relocate* operator and giant tour encoding

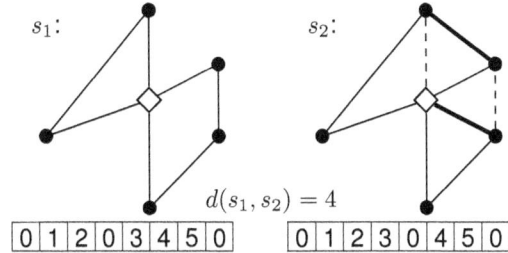

We implemented three algorithms to enumerate the solutions. 1) A permutation based algorithm lists all possible choices for the first points for K routes. All permutations of remaining points are then divided to the routes in all possible ways. 2) A matrix based algorithm that uses an integer interpreted as binary upper strict triangle matrix of decision variables. Increase by 1 gives a new solution. With constraints to the sum of a row/column we can prune entire branches of constraint breaking solutions. 3) An algorithm based on generation of all Hamiltonian paths after which visits to the depot are recursively added. A pseudocode of a subprocedure for the last method is given below.

Require: giant tour encoded solution s
Require: a, b are indexes of s: $a, b \in \{0, \dots, |s| - 1\}$
  $R_t \leftarrow \emptyset$
  if $b - a \geq 2$ **then**
    for $c = a$ **to** b **do**
      $t_{c-}, t_{c+} \leftarrow$ the tours of s before and after c
      $t_{b+} \leftarrow$ the tour of s after b
      $C_1 \leftarrow s[c] \neq 0$ **and** $s[c+1] \neq 0$
      $C_2 \leftarrow t_{c-}.first() < t_{c-}.last()$
      $C_3 \leftarrow t_{c+}.first() < t_{c+}.last()$
      $C_4 \leftarrow t_{c-}.first() < t_{c+}.first()$
      $C_5 \leftarrow |t_{b+}| > 0$ **and** $t_{c+}.first() < t_{b+}.first()$
      if C_1 **and** C_2 **and** C_3 **and** C_4 **and** C_5 **then**
        $s^* \leftarrow$ copy of s with visit to depot at c
        $R_t \leftarrow R_t \cup \{s^*\}$
        $R_t \leftarrow R_t \cup$ call algorithm recursively with
          $s' \leftarrow s^*, a' \leftarrow a, b' \leftarrow c+1$
        $R_t \leftarrow R_t \cup$ call algorithm recursively with
          $s' \leftarrow s^*, a' \leftarrow c+1, b' \leftarrow b$
      end if
    end for
  end if
  return R_t

We use giant tour encoding (Toth and Vigo, 2002) with 0 indicating the depot as illustrated in Figure 2. In addition we use rules that require the tours to be directed so that the first node has always smaller index than the last node (conditions C_2, C_3 of the procedure) and that tours are ordered according to their first nodes (conditions C_4, C_5). These enforce unique encoding of the solutions.

Require: $N \in \mathcal{N} \leftarrow$ Size of the CVRP instance
    $R \leftarrow \emptyset$
    for all $t \in$ Hamiltonian paths **do**
        if $t.first() < t.last()$ **then**
            $R \leftarrow R \cup \{t\}$
            $R \leftarrow R \cup$ call algorithm with
                $s' \leftarrow t, a' \leftarrow 1, b' \leftarrow N$
        end if
    end for
    return R

The algorithm described above is then used to generate all solutions of an asymmetric VRP using the subprocedure described earlier. Generation of Hamiltonian cycles can simply be done generating permutations of the request indices.

C. Multidimensional Scaling

MDS is a name for a family of techniques of dimensionality reduction for data analytics and visualization. Input is proximity data in the form of a dissimilarity matrix $D^{N \times N}$ of high-dimensional coordinate points and the output embeddings of these points into a lower dimensional space $\mathcal{R}^d$. Several methods exist, but in this work we use the popular SMACOF stress majorization algorithm (Borg and Groenen, 2005). As the minimization target we used the Kurskal stress function:

$$\sigma(X, D) = \sqrt{\sum_{i<j} (d_{ij} - \hat{d}_{ij})^2} \qquad (6)$$

We used the implementation from Orange Data Mining library[1] with Torgerson initialization (Borg and Groenen, 2005) that gives an analytic solution to the MDS. The SMACOF was terminated after 100 iterations or when termination condition $\sigma < 0.001$ was met.

D. Drawing the Visualization

To draw the visualization we used Mayavi2[2] library for scientific 3D data visualization. We used Delaunay 2D triangulation to create a mesh which was then plotted with Mayavi surface pipeline. LS algorithm trace is drawn using 3D parabolas as "jumps" from solution to another.

The accuracy of the visual representation can be verified using fitness landscape analysis measures. This idea is one the major contributions of this paper. We want to examine if important solution space features, such as the autocorrelation length λ of a random walk (RW) (Weinberger, 1990), number of local optima $\#LO$ discovered with random initialization and LS, and distribution of the aforementioned local optima $(\bar{d}(LO_i, LO_i^*), \sigma_d)$, are preserved. To do this we need to define a LS operator in the visualization space. In this work we chose to use k-nearest neighbour search, where k is the length giant tour encoding of the current solution.

---

[1] http://orange.biolab.si/
[2] http://code.enthought.com/projects/mayavi/

V. RESULTS

The described generation method is capable of producing around 220 000 solutions per second on a 2.13GHz Intel Core 2 Duo workstation. For any practical purposes this is enough.

TABLE I: Number of solutions with different n

n	3	4	5	6	7	8	9	10
$|R_{CVRP}|$	7	34	206	1486	12412	117692	1248004	14625856
$|R_{TSP}|$	1	3	12	60	360	2520	20160	181440

Because all three implemented algorithms produced the same number of solutions at least up to $n = 10$ we feel confident that all VRP solutions are enumerated. Enumeration with different number of customers yields a sequence shown in Table I. To allow comparison, we have also listed the number of possible solutions for a corresponding TSP expressed with $|R_{TSP}| = (n-1)!/2$.

Our empirical study of the number of solutions to symmetric CVRP leads to Formula (7) (Wilf, 2012). We have verified it produces the correct number of solutions least up to $N = 13$. By presenting the result, we would like to open discussion of the implications of knowing the exact number of solutions.

$$|R_{CVRP}| = \sum_{k=0}^{n} |s_1(n, k)| \, b(k), \quad \text{where} \qquad (7)$$

$$s_1(n, m) = \sum_{k=0}^{n-m} \binom{n-1+k}{n-m+k} \binom{2n-m}{n-m-k} s_2(n-m+k, k) \qquad (8)$$

$$s_2(n, m) = \sum_{k=0}^{m} \frac{\binom{m}{k} k^n}{m! (-1)^{(k-m)}} \qquad (9)$$

$$b(n) = \sum_{k=0}^{n} \sum_{i=0}^{k} \frac{(-1)^i (k-2i)^n \binom{k}{i}}{2^k k!}, \quad b(0) = 1 \qquad (10)$$

In order to examine the suitability of the distance measures we compare MDS stress. As a reference we use fitting of a uniformly random point cloud inside a unit 3D ball down to 2D coordinates. Reference provides us an intuition of how "tight" it gets when fitting CVRP solutions into 2D. The results are presented in Table II. Dice and Manhattan seem to be the ones showing stable behaviour and least stress.

TABLE II: MDS stress comparison

	CVRP				3D ball
$|R|$	Dice	Manhattan	Jaccard	Yule	euclidean
7	0.008	**0.004**	0.010	0.027	0.003
34	**0.019**	0.021	0.032	0.037	0.008
206	**0.035**	0.039	0.042	0.056	0.010
1486	0.049	0.048	0.073	**0.043**	0.011

In Figure 3 we present visualizations of randomly generated TSP and CVRP instances produced with our method. In addition, the rightmost plot visualizes a search trajectory of Clarke-Wright construction heuristic followed by few iterations of 2-opt. Our visualization tool allows interactive examination with freeform rotation around the visualizations[3].

---

[3] http://youtu.be/LW4DnTHpXoE
http://youtu.be/i6JsS6VjsEU
http://youtu.be/y53Qh2kC0Xo

(a) TSP, $n = 6$

(b) CVRP, $n = 6$

(c) CVRP LS, $| \cup N(s_i)| = 1218$

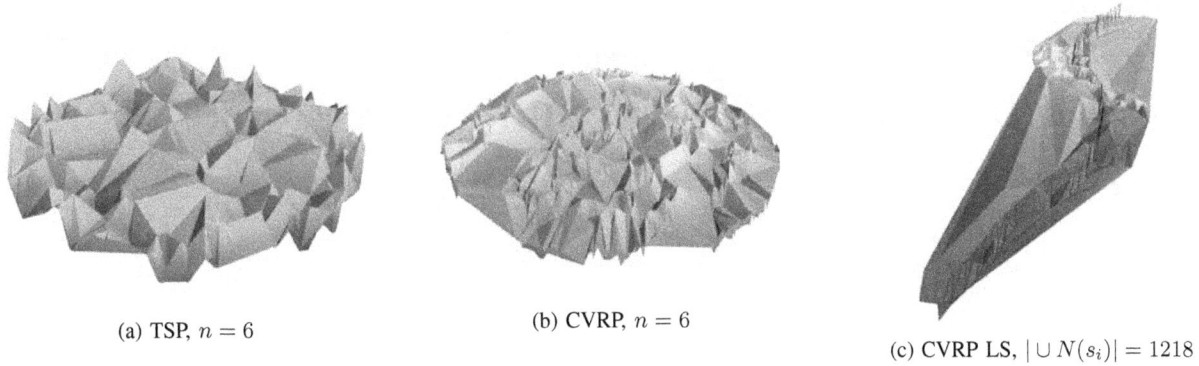

Fig. 3: Example visualizations produced using the described method

TABLE III: Perseverance of fitness landscape features

Measure	$CVRP_{n=4}$	$CVRP_{n=5}$	$CVRP_{n=6}$
λ	0.22	0.70	0.67
$\#LO$	0.73	0.24	0.01
$\bar{d}\,(\sigma_d)$	0.78 (0.53)	0.26 (0.32)	0.58 (0.59)

We calculated solution space metrics for 20 random CVRPs and compared them with the ones for corresponding visualizations. Each RW and LS with maximum length of 500 steps was repeated 100 times. Finally the results were averaged for reliability. Table III shows correlations between these results. It becomes clear that the accuracy of the described method drops on transition from $n = 5$ to $n = 6$. Especially the placement of local optima in a topology persevering way becomes hard. Luckily, the problem is smaller with neighbourhood plots as there is more room for the MDS to work. In addition, the visualization of neighbourhoods is more convenient in algorithm development as it allows e.g. plotting of different LS strategies into one visualization for comparison or troubleshooting.

VI. CONCLUSIONS AND FUTURE TOPICS

We have presented a technique capable of visualizing vehicle routing problem solution landscapes. Because of this technique, we were able to visually observe the nature of the vehicle routing problem instance landscape for the first time. We also examined the accuracy of our method. We have shown that the properties and possibilities make the presented technique a viable tool for routing algorithm designers, especially when dealing with local search neighbourhoods. We also presented the first closed form function for the number of solutions for symmetric CVRPs.

Further research might be extended the comparison to recent routing specific distance measures (Løkketangen et al., 2012). Also, accelerated methods such as Split-and-Combine MDS or Landmark MDS might be needed to visualize larger solution spaces and search algorithm neighbourhoods. Another bottleneck was the handling of big distance matrices. Experimentation with approximation and compression techniques could lead to faster calculations and reduced memory footprint. Our aim is to create an online gallery of problem variants (CVRP, TSP, VRPTW, PDP etc.) and search trajectories of different popular heuristic operators and metaheuristics. Also the source code of the tool will be published with the gallery.

REFERENCES

S. Halim and H. C. Lau, "Tuning tabu search strategies via visual diagnosis," in *Meta-Heuristics: Progress as Complex Systems Optimization.* Kluwer, 2007, pp. 365–388.

P. Toth and D. Vigo, *The vehicle routing problem.* SIAM, 2002.

E. Weinberger, "Correlated and uncorrelated fitness landscapes and how to tell the difference," *Biological Cybernetics*, vol. 63, no. 5, pp. 325–336, Sep. 1990.

C. Fonlupt, D. Robilliard, P. Preux, and E.-G. Talbi, "Fitness landscapes and performance of meta-heuristics," in *Meta-Heuristics: Advances and Trends in Local Search Paradigms for Optimization.* Kluwer, 1999, pp. 257–268.

M. Kubiak, "Distance measures and fitness-distance analysis for the capacitated vehicle routing problem," in *Metaheuristics: Pregress in Complex Systems Optimization.* Springer, 2007, pp. 345–364.

Z. J. Czech, "Statistical measures of a fitness landscape for VRP," in *Proceedings of IPDPS.* IEEE, 2008, pp. 1–8.

E. Pitzer, M. Affenzeller, A. Beham, and S. Wagner, "Comprehensive and automatic fitness landscape analysis using heuristiclab," in *EUROCAST'11.* Berlin, Heidelberg: Springer-Verlag, 2012, pp. 424–431.

F. Mascia and M. Brunato, "Techniques and tools for search landscape visualization and analysis," in *Proceedings of Stochastic Local Search 2009, Brussels, Belgium*, ser. LNCS, T. Stützle, M. Birattari, and H. Hoos, Eds., vol. 5752. Springer Berlin / Heidelberg, 2010, pp. 92–104.

M. Brunato and R. Battiti, "Grapheur: a software architecture for reactive and interactive optimization," in *LION'10.* Berlin, Heidelberg: Springer-Verlag, 2010, pp. 232–246.

S. S. Choi, S. H. Cha, and C. Tappert, "A Survey of Binary Similarity and Distance Measures," *Journal on Systemics, Cybernetics and Informatics*, vol. 8, no. 1, pp. 43–48, 2010.

I. Borg and P. Groenen, *Modern Multidimensional Scaling: Theory and Applications.* Springer, 2005.

H. S. Wilf. (2012) A011800: number of labeled forests of n nodes each component of which is a path. [Online]. Available: http://oeis.org/A011800

A. Løkketangen, J. Oppen, J. Oyola, and D. L. Woodruff, "An Attribute Based Similarity Function for VRP Decision Support," *Decision Making in Manufacturing and Services*, vol. 6, no. 2, pp. 65–83, 2012.

Computational Intelligence – the Key to Optimizing Workforce Management and Professional Sports League Schedules

Kimmo Nurmi
Satakunta University of Applied Sciences
Pori, Finland
cimmo.nurmi@samk.fi

Jari Kyngäs
Satakunta University of Applied Sciences
Pori, Finland
jari.kyngas@samk.fi

Nico Kyngäs
Satakunta University of Applied Sciences
Pori, Finland
nico.kyngas@samk.fi

Abstract— **Computational intelligence is a set of nature-inspired computational methods to address highly complicated real-world problems. Examples of such methods include neural networks, fuzzy logic and genetic algorithms. This paper presents our computational intelligence method called the PEAST algorithm. The algorithm has been used to solve school timetabling problems, many different workforce management problems and several sports scheduling problems. In this paper we summarize our work with real-world workforce scheduling and professional sports league scheduling. The algorithm has been integrated into market-leading workforce management software in Finland and it is used to generate the schedule for the Finnish Major Ice Hockey League.**

Keywords—computational intelligence; the PEAST algorithm; workforce scheduling; sports scheduling; real-world scheduling

I. INTRODUCTION

Computational intelligence (CI) has a fairly broad definition. We consider CI as a set of nature-inspired computational methods to address highly complicated real-world problems. These methods use ideas from their biological counterparts and are related in particular to artificial learning. Examples of such methods include neural networks [1], fuzzy logic [2] and genetic algorithms [3].

Computational intelligence is particularly useful in solving highly complex combinatorial optimization problems, such as workforce scheduling [4] and sports scheduling [5]. These scheduling problems have gained much attention among the researchers in the last decade. One important reason for this is that computers have evolved to such a state that they are able to calculate complicated and time-consuming mathematics in a reasonable amount of time.

Another reason is that the requirements and constraints of real-world problems have become more complicated which makes it impossible to produce solutions manually. Public institutions and private companies have become more aware of the possibilities for decision support technologies, and they no longer want to create the schedules manually. One further significant benefit of automating scheduling processes is the considerable amount of time saved by the administrative staff involved.

Section II describes our computational intelligence method, the PEAST algorithm. The most important components of the algorithm are discussed in more detail. These components are the key to successful real-world use. In Section III we summarize our work with the Finnish Major Ice Hockey League. We also report our computational results in generating the schedule for the 2013-2014 season. Section IV summarizes our work with workforce scheduling. We report two computational results in scheduling workforce in Finnish companies. Section V presents conclusions and some of our future work.

II. THE PEAST ALGORITHM

The most successful CI methods do not solely use neural networks, fuzzy logic or genetic algorithms. They most likely use mixed local search and population-based methods. A local search method is defined by

1) A neighborhood structure, which is a mechanism to obtain a new set of neighbor solutions by applying a small perturbation to a given solution.

2) A method of moving from one solution to another.

3) Parameters of the method.

The main difficulty for a local search algorithm is

1) to explore promising areas in the search space to a sufficient extent, while at the same time

2) to avoid staying stuck in these areas too long and

3) to escape from these local optima in some systematic way.

Population-based methods use a population of solutions in each iteration. The outcome of an iteration is also a population of solutions. Population-based methods are a good way to escape from local optima.

Osman and Laporte [6] define metaheuristic as follows: "A metaheuristic is an iterative generation process which guides a subordinate heuristic by combining intelligent concepts for exploring and exploiting the search space. Furthermore, learning strategies are used to structure information in order to efficiently find near-optimal solutions".

Blum and Roli [7] classify metaheuristics based on five criteria:

1) Nature-inspired algorithms are those that are inspired by some phenomenon in the nature.

2) Population based algorithms work with many simultaneous solution candidates. This may lead to high-quality solutions but the computing time increases very quickly.

3) Dynamic objective function means that the algorithm modifies the weights of the multiobjective function during the search. This can be of great help when the importance and difficulty of the objectives are not known or cannot even be guessed.

4) One vs. various neighborhood structures. Using various neighborhood structures usually makes the search space larger. An algorithm can search one neighborhood for some time and swap into another if good solutions are not found.

5) Memory usage means that an algorithm saves the search history while running. This helps the evaluation of search locations – if the search location has been visited earlier the algorithm can use that information.

Metaheuristic methods guide the search process towards near-optimal solutions. The idea is to efficiently explore the search space with a combination of simple local search procedures and/or complex learning processes. The methods are usually non-deterministic and they may include mechanisms to avoid getting trapped in confined areas of the search space (i.e. local minima).

Popular metaheuristic methods include ant colony optimization [8], ejection chains [9], genetic algorithms, hill-climbing, hyper-heuristics [10], idwalk [11], iterated local search, memetic algorithms [12], particle swarm optimization [13], simulated annealing [14], tabu search [15] and variable neighborhood search [16].

The usefulness of an algorithm depends on several criteria. The two most important ones are the quality of the generated solutions and the algorithmic power of the algorithm (i.e. its efficiency and effectiveness). Other important criteria include flexibility, extensibility and learning capabilities. We can steadily note that our PEAST algorithm realizes these criteria. It has been used to solve several real-world scheduling problems and it is in industrial use. The first version of the algorithm was used to solve school timetabling problems [17].

The later versions of the algorithm have been used to solve sports scheduling problems (see e.g. [18] and [19]) and workforce scheduling problems (see e.g. [20] and [21]).

The PEAST algorithm is a population-based local search method. It is very difficult to classify the algorithm as one of the earlier mentioned metaheuristics. The basic hill-climbing step is extended to generate a sequence of moves in one step, leading from one solution to another as is done in the ejection chain method. The algorithm avoids staying stuck (i.e. the objective function value does not improve for some predefined number of generations) in the same solutions using tabu search and the refined simulated annealing method. The algorithm belongs to population-based methods in that it uses a population of solutions. This enables it to explore promising areas in the search space to a sufficient extent. To escape from local optima the algorithm shuffles the current solution mimicking a hyper-heuristic mechanism. The acronym PEAST stems from the methods used: Population, Ejection, Annealing, Shuffling and Tabu. The pseudo-code of the algorithm is given in Figure 1.

Set the iteration limit t, cloning interval c, shuffling interval s, ADAGEN update interval a and the population size n
Generate a random initial population of schedules S_i for $1 <= i <= n$
Set $best\_sol = null$, $round = 1$
WHILE $round \leq t$
   $index = 1$
   WHILE $index++ <= n$
      Apply GHCM heuristic to schedule S_{index} to get a new schedule
      IF $Cost(S_{index}) < Cost(best\_sol)$ THEN Set $best\_sol = S_{index}$
   END WHILE
   Update simulated annealing framework
   IF $round \equiv 0$ (mod a) THEN **Update the ADAGEN framework**
   IF $round \equiv 0$ (mod s) THEN **Apply shuffling operators**
   IF $round \equiv 0$ (mod c) THEN **Replace the worst schedule with the best one**
   Set $round = round + 1$
END WHILE
Output $best\_sol$

Fig. 1. The pseudo-code of the PEAST algorithm.

The GHCM heuristic is based on similar ideas to ejection chains [19] and the Lin-Kernighan procedures [22]. The basic hill-climbing step is extended to generate a sequence of moves in one step, leading from one solution candidate to another. The GHCM heurictis moves an object, o_1, from its old position, p_1, to a new position, p_2, and then moves another object, o_2, from position p_2 to a new position, p_3, and so on, ending up with a sequence of moves.

Picture the positions as cells as shown in Figure 2. The initial object selection is random. The cell that receives the object is selected by considering all the possible cells and selecting the one that causes the least increase in the objective function when only considering the relocation cost. Then, another object from that cell is selected by considering all the objects in that cell and picking the one for which the removal causes the biggest decrease in the objective function when only considering the removal cost. Next, a new cell for that object is selected, and so on. The sequence of moves stops if the last move causes an increase in the objective function value and if the value is larger than that of the previous non-improving move, or if the maximum number of moves is reached. Then, a new sequence of moves is started. The maximum number of moves in the sequence is ten.

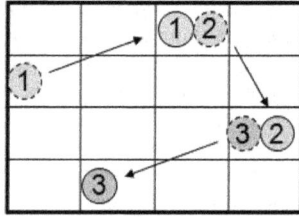

Fig. 2. A sequence of moves in the GHCM heuristic.

We improve the GHCM heuristic by introducing a tabu list which prevents reverse order moves in the same sequence of moves. i.e. if we move an object o from position p_1 to position p_2, we do not allow o to be moved back to position p_1 before a new sequence of moves begins.

Simulated annealing [14] is a celebrated local search heuristic which avoids staying stuck in the promising search areas too long. We use a simulated annealing refinement to decide whether or not to commit to a sequence of moves in the GHCM heuristic. This refinement is different from the standard simulated annealing. It is used in a three-fold manner. Firstly, when choosing an object to be moved from a cell, a random object is chosen with probability $\exp(-1/T_k)$ instead of choosing the least fit object. Here T_k is the temperature at step k. Secondly, when choosing the cell where to move the object, a random cell is chosen with probability $\exp(-1/T_k)$ instead of choosing the fittest cell. Lastly, when the sequence of moves is cut short (i.e. a worsening move is made, and it worsens the solution more than the previous worsening move did), the whole sequence will still be committed with probability $\exp(-\Delta f/T_k)$ instead of rolling back to the best position (i.e. the position at which the objective function value is the lowest) of the sequence. Here Δf is an increase in the cost function.

We calculate the initial temperature T_0 by

$$T_0 = 1/\log\left(X_0^{-1}\right)$$

where X_0 is the degree to which we want to accept an increase in the cost function (we use a value of 0.75). The exponential cooling scheme is used to decrement the temperature:

$$T_k = \alpha T_{k-1},$$

where α is usually chosen between 0.8 and 0.995. We stop the cooling at some predefined temperature. Therefore, after a certain number of iterations, m, we continue to accept an increase in the cost function with some constant probability, p. Using the initial temperature given above and the exponential cooling scheme, we can calculate the value

$$\alpha = \left(-1/\left(T_0 \log p\right)\right)^{1/m}$$

We choose m equal to the maximum number of iterations and p equal to 0.0015.

For most PEAST applications we introduce a number of shuffling operators – simple heuristics used to perturb a solution into a potentially worse solution in order to escape from local optima – that are called upon according to some rule. We call the operator every $k/20$th iteration of the algorithm, where k equals the maximum number of iterations with no improvement to the cost function. The idea of shuffling is the same as in hyperheuristics [10] but the other way around. A hyperheuristic is a mechanism that chooses a heuristic from a set of simple heuristics, applies it to the current solution to get a better solution, then chooses another heuristic and applies it, and continues this iterative cycle until the termination criterion is satisfied. We introduce a number of simple heuristics that are used to worsen the current solution instead of improving it.

No crossover operators are applied to the population of schedules. Every c iterations the least fit individual is replaced with a clone of the fittest individual. This operation is completely irrespective of the globally fittest schedule (*best_sol* in Figure 1) found.

The PEAST algorithm uses ADAGEN, the adaptive genetic penalty method introduced in [17]. A traditional penalty method assigns positive weights (penalties) to the soft constraints and sums the violation scores to the hard constraint values to get a single value to be optimized. The ADAGEN method assigns dynamic weights to the hard constraints based on the constant weights assigned to the soft constraints. The soft constraints are assigned fixed weights according to their significance. This means that we are searching for a solution that minimizes the (penalty) function

$$\sum_i \alpha_i f_i(x) + \sum_i c_i g_i(x),$$

where

α_i = a dynamically adjusted weight for hard constraint i
$f_i(x)$ = cost of violations of hard constraint i
c_i = a fixed weight for soft constraint i
$g_i(x)$ = cost of violations of soft constraint i.

The hard constraint weights are updated every kth generation using the method given in [12].

The PEAST algorithm uses random initial solutions. We have found no evidence that a sophisticated initial solution improves results. On the contrary, random initial solutions seem to yield superior or at least as good results.

III. PROFESSIONAL SPORTS LEAGUE SCHEDULING

In the past decades professional sports leagues have become big businesses; at the same time the quality of the schedules have become increasingly important. This is not surprising, since the schedule directly impacts the revenue of all involved parties. For instance, the number of spectators in the stadiums, and the traveling costs for the teams are influenced by the schedule. TV networks that pay for broadcasting rights want the most attractive games to be scheduled at commercially interesting times in return. Furthermore, a good schedule can make a tournament more interesting for the media and the fans, and fairer for the teams. An excellent overview of sports scheduling can be found in [23] and an annotated bibliography in [24].

In a sports tournament, *n* teams play against each other over a period of time according to a given timetable. The teams belong to a league, which organizes *games* between the teams. Each game consists of an ordered pair of teams, denoted (i, j), where team *i* plays *at home* - that is, uses its own venue (stadium) for a game - and team *j* plays *away*. Games are scheduled in *rounds*, which are played on given days. A *schedule* consists of games assigned to rounds. If a team plays two home or two away games in two consecutive rounds, it is said to have a *break*. In general, for reasons of fairness, breaks are to be avoided. However, a team can prefer to have two or more consecutive away games if its stadium is located far from the opponent's venues, and the venues of these opponents are close to each other. A series of consecutive away games is called an *away tour*.

In a *round robin tournament* each team plays against each other team a fixed number of times. Most sports leagues play a double round robin tournament (*2RR*), where the teams meet twice (once at home, once away), but quadruple round robin tournaments (*4RR*) are also quite common.

Sports scheduling involves three main problems. First, the problem of finding a schedule with the *minimum number of breaks* is the easiest one. De Werra [25] has presented an efficient algorithm to compute a minimum break schedule for a *1RR*. If *n* is even, it is always possible to construct a schedule with $n - 2$ breaks.

Second, the problem of finding a schedule that minimizes the travel distances is called the *Traveling Tournament Problem* (TTP) [26]. In TTP the teams do not return home after each away game but instead travel from one away game to the next. However, excessively long away trips as well as home stands should be avoided.

Third, most professional sports leagues introduce many additional requirements in addition to minimizing breaks and travel distances. The problem of finding a schedule which satisfies given constraints is called the *Constrained Sports Scheduling Problem* (CSSP). The goal is to find a feasible solution that is the most acceptable for the sports league owner - that is, a solution that has no hard constraint violations and that minimizes the weighted sum of the soft constraint violations.

Scheduling the Finnish major ice hockey league is an example of a CSSP. It is very important to minimize the number of breaks. The fans do not like long periods without home games, consecutive home games reduce gate receipts and long sequences of home or away games might influence the team's current position in the tournament. It is also very important to minimize the travel distances. Some of the teams do not return home after each away game but instead travel

from one away game to the next. There are also around a dozen more other criteria that must be optimized. Next we describe the problem in more detail.

Ice hockey is the biggest sport in Finland, both in terms of revenue and the number of spectators. The spectator average per game for the 2012-2013 season was 5213. In the Saturday rounds one percent of the Finnish population (age 15-70) attended the games in the ice hockey arenas.

The Finnish major ice hockey league has 14 teams. Seven of the teams in the league are located in big cities (over 100,000 citizens) and the rest in smaller cities. One team is quite a long way up north, two are located in the east and the rest in the south.

The basis of the regular season is a quadruple round robin tournament resulting in 52 games for each team. In addition, the teams are divided into two groups of seven teams in order to get a few more games to play. The teams in the groups are selected based on fairness, i.e. the strengths of the teams are most likely to be equal. These teams play a single round robin tournament resulting in 6 games. The home teams of the games are decided so that in two consecutive seasons each team has exactly one home game against every other team in the group. Finally, the so-called "January leveling" adds two extra games for each team.

The quadruple round robin, six group games and two extra games per team total to 60 games for each team and 420 games overall. The standard game days are Tuesday, Friday and Saturday. The schedule should maximize the number of games on Fridays and on Saturdays in order to maximize the revenue. For the same reason consecutive home games are not allowed on Fridays and Saturdays. Furthermore, due to the travel distances between some venues, certain combinations of a Friday home team playing a Saturday away game against the given team are not allowed.

Every team should play in the two rounds before and the two rounds after New Year's Eve and adhere to the traveling rules. The schedule should include a weekend when seven pairs of teams play against each other on consecutive rounds on Friday and on Saturday. These match-up games are called "back-to-back games". Furthermore, local rivals should play as many games as possible against each other in the first two rounds and the number of Friday and Saturday games between them should be maximized.

The traveling distances between some of the venues require some teams to make away tours. That is, they should play away either on Tuesday and on Thursday or on Thursday and on Saturday.

Two teams cannot play at home on the same day because they share a venue. Two other teams cannot play at home on the same day because they share the same (businessmen) spectators. Some of the teams cannot play at home in certain days because their venues are in use for some other event. Finally, in the last two rounds each team should play exactly one home game. The detailed constraint model of the problem can be found in [27].

We believe that scheduling the Finnish major ice hockey league is one of the most difficult sports scheduling problems because it combines break minimization and traveling distance restrictions with dozens of constraints that must be satisfied. We have used the PEAST algorithm and its predecessors to schedule the league since the 2008-2009 season. Table I shows the increase in the number of spectators in the last five seasons.

TABLE I.   THE NUMBER OF SPECTATORS IN THE FINNISH MAJOR ICE HOCKEY LEAGUE IN THE LAST EIGHT SEASONS

2005-2006	1 958 843
2006-2007	1 943 312
2007-2008	1 964 626
2008-2009	1 997 114
2009-2010	2 015 080
2010-2011	2 036 915
2011-2012	2 145 462
2012-2013	2 189 350

TABLE II.   THE BEST SCHEDULE FOUND

C01	Hard	There are at most 90 rounds available for the tournament	0
C04	Hard	A team cannot play at home in the given round (43 cases)	0
C07	Hard	Two pairs of teams cannot play at home in the same round (2 cases)	0
C08	Hard	A team cannot play at home on two consecutive calendar days	0
C12	Hard	A break cannot occur in the second and last rounds	0
C41	Hard	"Back-to-back games"	0
C09	Soft 10	Number of away tours not scheduled	0
C13	Soft 10	Cases when a team has more than two consecutive home games	0
C14	Soft 3	Cases when a team has more than two consecutive away games	0
C15	Soft 5	The schedule should have at most 140 breaks	0
C19	Soft 5	Cases when there are less than five rounds between two games with the same opponents	1
C22	Soft 1	Cases when two teams play against each other in series of HHAA, AAHH, HAAAH or AHHHA	4
C23	Soft 1	Cases when a team does not play its home games in given weekdays (team's wishes)	2
C26	Soft 1	Cases when the difference between the number of played home and away games for each team is more than two in any stage of the tournament	1
C27	Soft 3	Cases when the difference in the number of played home games between the teams is more than two in any stage of the tournament	0
C37	Soft 4	Cases when a team's travel distance in Friday and Saturday games is too long	0
C37	Soft 4	The compact rounds around New Year's Eve should be considered	0
C38	Soft 10	The games between local rivals should be considered	0
C39	Soft 10	Cases when the games between local rivals are not in Fridays or in Saturdays	0
C40	Soft 10	The schedule should include at least 200 games played either on Friday or on Saturday	0

The standard game days used to be Tuesday, Thursday and Saturday. The league changed Thursdays to Fridays to get more spectators. Friday games have had about 10% more spectators. However, this has complicated the scheduling.

Minimizing the number of breaks is very important because it is likely that having two consecutive home games on Thursday and on Saturday decreases the number of spectators. In the last ten seasons the number of spectators on Thursday has decreased by 3.5% and on Saturday by 1.9%.

The process of scheduling the league takes about two months. First, we discuss the possible improvements to the format with the league's competition manager. Then, the format is accepted by the team CEOs. Next, all the restrictions, requirements and requests by the teams are gathered. Finally, the importance (penalty value) of the constraints is decided. We run the PEAST algorithm for one week several times and choose the best schedule. The league accepts the final schedule.

Table II shows the best solution found for the 2013-2014 season. The last column indicates the number of hard/soft constraint violations. In [28] we presented a collection of typical constraints that are representative of many scheduling scenarios in sports scheduling. In Table II we refer to this constraint classification. The solution has no hard constraint violations and the penalty value for the soft constraint violations is 12. The schedule has 89 rounds (C01), 139 breaks (C15) and 220 games played either on Friday or on Saturday (C40). This was by far the most difficult schedule to generate compared to the earlier seasons.

IV. WORKFORCE SCHEDULING

Workforce scheduling is a difficult and time consuming task that many companies must solve, especially when employees are working on shifts or on irregular working days. The workforce scheduling problem has a fairly broad definition. A good overview of workforce scheduling is published by Ernst et al. [29].

There are hundreds of workforce scheduling solutions commercially available and in widespread use. However, most of the commercial products do not include any computational intelligence. Most of the workforce scheduling research concentrates on nurse rostering. However, the recent interest of the academic workforce scheduling community has somewhat shifted to other lines of businesses. Some of these are retail sector, transportation sector and contact centers.

Figure 3 shows the phases of the workforce scheduling process. The process can be divided to preprocessing, main scheduling and supplemental phases.

A. Preprocessing phases

The preprocessing phases of the workforce scheduling process, workload prediction and preference scheduling, are the foundation upon which the scheduling phases are built. They may involve identifying both the needs of the employer and customers and the attributes (preferences, skills etc.) of the employees, and determining staffing requirements. This is the point in the workforce scheduling process where historical data and the schedules of previous planning horizons are most useful.

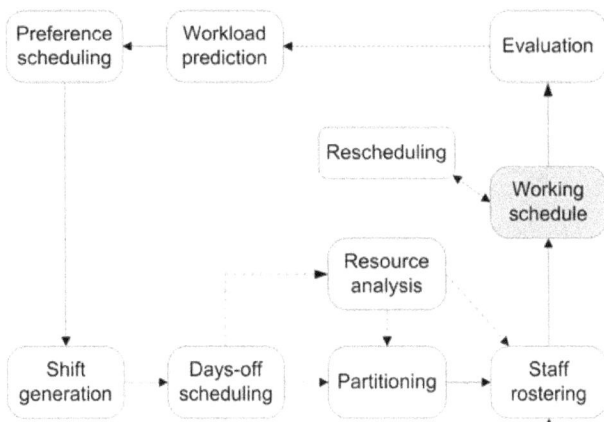

Fig. 3. The workforce scheduling process.

Workload prediction, also referred to as demand forecasting or demand modeling, is the process of determining the staffing levels - that is, how many employees are needed for each timeslot in the planning horizon. The staffing is preceded by actual workload prediction or workload determination based on static workload constraints given by the company, depending on the situation. In *preference scheduling*, each employee gives a list of preferences and attempts are made to fulfill them as well as possible. The employees' preferences are often considered in the days-off scheduling and staff rostering phases, but may also be considered during shift generation.

The nature of determining the amount and type of work to be done at any given time during the next planning horizon depends greatly on the nature of the job. If the workload is uncertain then some form of workload prediction is called for [29]. Some examples of this are the calls incoming to a call center or the customer influx to a hospital. We have simulated the randomly distributed workload based on historical data and statistical analysis to find a suitable working structure (i.e. how many and what kinds of employees are needed) over time. Computationally this approach is much more intensive than methods based on queuing theory. However, it has the benefit of being applicable to almost any real-world situation. If the workload is static, no forecasting is necessary. For example, a bus transit company might be under a strict contract to drive completely pre-assigned bus lines.

It is crucial for a workforce management system to allow the employees to affect their own schedules. In general it improves employee satisfaction. This in turn reduces sick leaves and improves the efficiency of the employees, which means more profit for the employer. Hence we use an easy-to-use user interface that allows the employees to input their preferences into the workforce management system. This eases the organizational workload of the personnel manager. A measure of fairness is incorporated via limiting the number and type of different wishes that can be expressed per employee. Preferences can be considered at the days-off scheduling and staff rostering phases.

B. Main scheduling phases

The main scheduling phases, shift generation, days-off scheduling and staff rostering, can be solved using computational intelligence. Computational workforce scheduling is key to increased productivity, quality of service, customer satisfaction and employee satisfaction. Other advantages include reduced planning time, reduced payroll expenses and ensured regulatory compliance.

Shift generation is the process of determining the shift structure, along with the activities to be carried out in particular shifts and the competences required for different shifts. Shift generation transforms the determined workload into shifts. This includes deciding break times when applicable. Shift generation is essential especially in cases where the workload is not static. In other cases companies often want to hold on to their own established shift building methods.

A basic shift generation problem includes a variable number of activities for each task in each timeslot. Some tasks are not time-dependent; instead, there may be a daily quota to be fulfilled. Activities may require competences. The most important optimization target is to match the shifts to the workload as accurately as possible. In our solutions we create the shifts for each day separately, each shift corresponding to a single employee's competences and preferences. We do not minimize the number of different shifts.

Days-off scheduling decides the rest days and the working days of the employees. It is based on the result of the shift generation: for each day a set of suitable employees must be available to carry out the shifts. The coverage requirement ensures that there are a sufficient number of employees on duty at all times. The regulatory requirements ensure that the employee's work contract and government regulations are respected. The personnel's requests are very important and should be met as well as possible; this leads to greater staff satisfaction and commitment, and reduces staff turnover.

The final optimization phase of the workforce scheduling process is *staff rostering*, during which the shifts are assigned to the employees. The length of the planning horizon for this phase is usually between two and six weeks. The preferences of the employees are usually given a relatively large weight. The most important constraints are usually resting times and certain competences, since these are often laid down by the collective labor agreements and government regulations.

C. Supplemental phases

To see if there will be any chance of succeeding at matching the workforce with the shifts while adhering to the given constraints, a *resource analysis* is run on the data. In addition to helping the personnel manager see the problem with the data, it may help in convincing the management level that the current practices and processes of generating the schedules are simply untenable. We have developed a statistical tool for this.

Some real-world datasets are huge. They may consist of hundreds of employees with a corresponding number of jobs. In these cases it is probably computationally impossible to try to roster the whole set of employees at once. We use the PEAST algorithm to intelligently *partition* the data.

First the employees are partitioned according to their average length of working day and possibly some other criteria into a number of approximately equal-sized groups. Then the jobs are partitioned into groups so that each group of jobs corresponds to one group of employees, i.e. so that it should be relatively easy to assign the jobs of one group to the corresponding employee group. The assignments can be done in parallel.

D. Two computational results

According to our experience, the best action plan for real-world workforce scheduling research is to cooperate both with a problem owner and with a third-party vendor. In addition, an academic should consider not to work with user interfaces, financial management links, customer reports, help desks, etc. Instead, one should concentrate on modeling issues and algorithmic power.

It is difficult to incorporate the experience and expertise of the personnel managers into a workforce scheduling system. Personnel managers often have extremely valuable knowledge, experience and detailed understanding of their specific staffing problem, which will vary from company to company. To formalize this knowledge into constraints is not an easy task.

We have used the PEAST algorithm to solve real-world shift generation problems in an international transportation and logistics company, in a Finnish contact center and in the biggest retail trade company. We next discuss the contact center case. Note that only few shift generation cases have been reported in the academic literature.

The contact center has 65 employees. The opening hours of the contact center are from 6am to 10pm for every business day. The length of a timeslot is 15 minutes. There are three tasks in total, denoted by A, B and C. The total workload (in timeslots) for task B is 594 and for task C 570. Only task A has staff demand that is dependent on the time of day as shown in Figure 4. The distribution of the employees' competences for the tasks and allowed/preferred shift lengths are given in Table III.

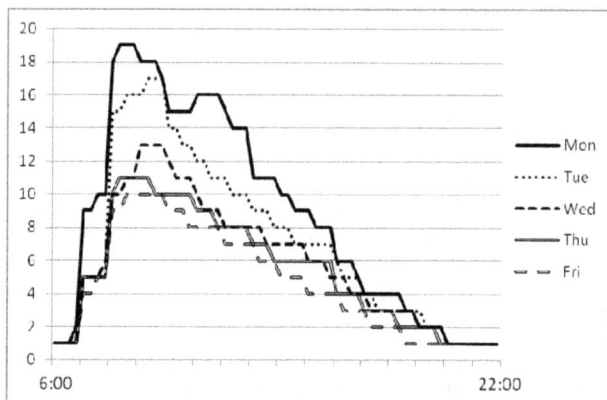

Fig. 4. Staff demand for task A by day and time of day.

TABLE III. THE DISTRIBUTION OF THE EMPLOYEE'S COMPETENCES AND ALLOWED/PREFERRED SHIFT LENGTHS

Competences	Allowed 8h	Preferred 8h, allowed 4-8h	No preference, allowed 4-8h
Only for A	0	13	6
Only for B	8	6	3
Only for C	4	6	8
For A and B	0	7	0
For A and C	0	3	1

Table IV shows the best solution found for the problem. For the hard and soft constraints of the problem we use the classification we presented in [30]. The solution has no hard constraint violations. The hard structure constraints (S1, S2) ensure that the shifts are physically feasible. The hard volume constraints (V5, V6) ensure that the employees can carry out the shifts. The hard placement constraints ensure that each shift has the correct number of suitable breaks (P3). Limiting the minimum stretch length to two hours (P5) and the number of task switches to less than one per shift (P6) are not inherently hard constraints, but they were chosen to be such to model their importance to the customer. The timeslot-based deviation from the staff demand (C1) was chosen to incur quadratic costs in order to both emphasize its importance and to balance the individual timeslots' deviations. The rest of the soft constraints are close to linear by nature.

TABLE IV. THE BEST SHIFT GENERATION SOLUTION FOUND

S1	Hard	No shift should contain timeslots with multiple types of activities	0
S2	Hard	No shift should contain gaps, i.e. timeslots with no activities	0
V5	Hard	The lengths of the shifts must match the employees' available hours	0
V6	Hard	The competences necessary to carry out the shifts must match the available workforce	0
P3	Hard	Each shift strictly shorter than 4 hours must contain one 15-minute break. Each shift at least 4 and strictly less than 6 hours in length must contain two 15-minute breaks. Each shift at least 6 hours in length must contain two 15-minute breaks and one 30-minute break.	0
P5	Hard	No task should appear in stretches shorter than two hours	0
P6	Hard	Each shift should contain at most one switch from one task to another	0
C1	Soft 1	The number of employees at each timeslot over the planning horizon must be exactly as given	32
C4	Soft 1	The total workload for each task must be carried out	2
V1	Soft 1	The number of shifts, i.e. the number of employees at work, must be minimized	286
V5	Soft 1	The lengths of the shifts must match the employees' preferred hours	2
P4	Soft 1	Within any shift, the breaks should be in the order short-long(-short) and spaced so that the distance between any adjacent breaks is equal	4
P7	Soft 1	Each switch between tasks should occur during a break	5

We have used the PEAST algorithm to solve real-world staff rostering problems in the biggest local bus company in the capital city of Finland, in the biggest nationwide transportation company in Finland and in one of the biggest hospitals in Finland. We next discuss the hospital case.

The intensive-care unit of the hospital has 130 employees. The employees must be assigned to shifts over the planning horizon of six weeks. The hard and soft constraints for the problem are given in Table V Note that the shifts and the days-off are optimized at the same time. The problem includes five characteristics that are not always present in the nurse rostering cases reported in the academic literature:

1) The number of nurses is over 100

2) The nurses are grouped in four categories based on their total working hours within the planning horizon (100%, 78.43%, 50% and 40% of the full-time work)

3) Some shifts last more than 14 hours and actually include two consecutive shifts

4) Some nurses should always work on the same shifts

5) The nurses' wishes for days-off and shifts cover as much as 50% of their total work.

TABLE V.   THE BEST STAFF ROSTERING SOLUTION FOUND

C1	Hard	An employee cannot be assigned to overlapping shifts	0
C2	Hard	A minimum number of employees with particular competences must be guaranteed for each shift	0
R5	Hard	The minimum time gap of rest time (11 hours) between two shifts must be respected	0
R7	Hard	Employees cannot work consecutively for more than 6 days	0
O1	Hard	An employee can only be assigned to a shift he/she has competence for	0
O5	Hard	An employee assigned to a shift type t1 must not be assigned to a shift type t2 on the following day	0
C4	Soft 2	A balanced number of surplus employees must be guaranteed in each working day	4
R1	Soft 10	The required number of working hours must be respected	3
R3	Soft 4	The required number of free weekends (both Saturday and Sunday free) within a planning horizon must be respected	16
E1	Soft 4	Single days-off should be avoided	124
E2	Soft 2	Single working days should be avoided	53
E3	Soft 2	The maximum length of consecutive days-off is three	0
E4	Soft 1	The number of single days-off and single working days between employees should not differ by more than 25%	5
E5	Soft 1	The number of different shift types between employees should not differ by more than 50%	46
E7	Soft 1	The number of different working weekdays between employees should not differ by more than 50%	127
E8	Soft 4	Assign or avoid a given shift type before or after a free period	0
P1	Soft 5	Assign or avoid assigning given employees to the same shifts	0
P2	Soft 6	Assign a requested day-on or avoid a requested day-off	**
P3	Soft 6	Assign a requested shift or avoid an unwanted shift	**

** more than 95% fulfilled

The best solution generated by the PEAST algorithm is shown in Table V. For the hard and soft constraints of the problem we use the classification we presented in [31]. The solution has no hard constraint violations. The PEAST algorithm was able to find a solution where all but one employee had exactly the required number of working hours. The algorithm also found a solution where more than 95% of all the employees' wishes were fulfilled even though those wishes covered as much as 50% of the employees' total work on average.

V.   CONCLUSIONS AND FUTURE WORK

We introduced our computational intelligence method called the PEAST algorithm and presented its most important components. We summarized our work with real-world workforce scheduling and professional sports league scheduling. We reported our main results in generating schedules for the Finnish Major Ice Hockey League and in scheduling workforce for Finnish companies. The PEAST algorithm has been integrated into market-leading workforce management software in Finland and it has been used to generate the schedule for the Finnish Major Ice Hockey League since 2008-2009 season.

Our future work concentrates on solving problems which include both location and vehicle routing in addition to workforce scheduling. Furthermore, we seek to extend our workforce scheduling research to new lines of businesses.

REFERENCES

[1]   R. Hecht-Nielsen, Neurocomputing, Addison Wesley, 1990.

[2]   V. Novák, I. Perfilieva and J. Močkoř, "Mathematical principles of fuzzy logic", the Springer International Series in Engineering and Computer Science, Vol. 517, 1999.

[3]   D.E. Goldberg, Genetic Algorithms in Search, Optimization and Machine Learning, Kluwer Academic Publishers, 1989.

[4]   E.I. Ásgeirsson, J. Kyngäs, K. Nurmi and M. Stølevik, "A Framework for Implementation-Oriented Staff Scheduling", in Proc of the 5th Multidisciplinary Int. Scheduling Conf.: Theory and Applications (MISTA), Phoenix, USA, 2011, pp. 308-321.

[5]   K. Nurmi, D. Goossens, T. Bartsch, F. Bonomo, D. Briskorn, G. Duran, J. Kyngäs, J. Marenco, CC. Ribeiro, FCR. Spieksma, S. Urrutia and R. Wolf-Yadlin, "A Framework for Scheduling Professional Sports Leagues", in Ao, Sio-Iong (ed.): IAENG Transactions on Engineering Technologies Volume 5, Springer, USA, 2010.

[6]   I.H. Osman and G. Laporte, "Metaheuristics: A bibliography", Annals of Operations Research 63, 1996, pp. 513-623.

[7]   C. Blum and A. Roli, "Metaheuristics in Combinatorial Optimization: Overview and Conceptual Comparison", ACM Computing Surveys 35(3), 2003, pp. 268-308.

[8]   A. Colorni, M. Dorigo adn V. Maniezzo, "Distributed Optimization by Ant Colonies", in F. J. Varela and P. Bourgine, editors, Towards a Practice of Autonomous Systems: Proceedings of the First European Conference on Artificial Life, 1992, pp. 134-142.

[9]   F. Glover, "New ejection chain and alternating path methods for traveling salesman problems", in Computer Science and Operations Research: New Developments in Their Interfaces, edited by Sharda, Balci and Zenios, 1992, pp. 449–509.

[10]   E.K. Burke, E. Hart, G. Kendall, J. Newall, P. Ross and S. Schulenburg, "Hyper-heuristics: An emerging direction in modern search technology", in Handbook of Metaheuristics, edited by F. Glover and G. Kochenberger, 2003, pp. 457–474.

[11] B. Neveu, G. Trombettoni and F. Glover, "Idwalk : A candidate list strategy with a simple diversification device", Lecture Notes in Computer Science 3258, 2004, pp. 423–437.

[12] P. Moscato, "On evolution, search, optimization, genetic algorithms and martial arts: Towards memetic algorithms", Technical Report 826, California Institute of Technology, Pasadena, California, 1989.

[13] R.C. Eberhart and J. Kennedy, "A New Optimizer Using Particle Swarm Theory", Proceedings of the Sixth International Symposium on Micromachine and Human Science, Nagoya, Japan, 1995, pp. 39-43.

[14] S. Kirkpatrick, C.D. Gelatt Jr and M.P. Vecchi, Optimization by Simulated Annealing, Science 220, 1983, pp. 671-680.

[15] F. Glover, C. McMillan and B. Novick, "Interactive Decision Software and Computer Graphics for Architectural and Space Planning", Annals of Operations Research 5, 1985, pp. 557-573.

[16] P. Hansen and C.N. Mladenovi, "Variable neighbourhood search: Principles and applications", European Journal of Operations Research 130, 2001, pp. 449–467.

[17] K. Nurmi, "Genetic Algorithms for Timetabling and Traveling Salesman Problems", Ph.D. dissertation, Dept. of Applied Math., University of Turku, Finland, 1998. Available: http://www.bit.spt.fi/cimmo.nurmi/

[18] J. Kyngäs and K. Nurmi, "Scheduling the Finnish Major Ice Hockey League", in Proc of the IEEE Symposium on Computational Intelligence in Scheduling, Nashville, USA, 2009.

[19] J. Kyngäs and K. Nurmi, "Scheduling the Finnish 1st Division Ice Hockey League", in Proc of the 22nd Florida Artificial Intelligence Research Society Conference, Florida, USA, 2009.

[20] N. Kyngäs, K. Nurmi, E.I. Ásgeirsson and J. Kyngäs, "Using the PEAST Algorithm to Roster Nurses in an Intensive-Care Unit in a Finnish Hospital", in Proc. of the 9th Conference on the Practice and Theory of Automated Timetabling (PATAT), Son, Norway, 2012.

[21] N. Kyngäs, K. Nurmi and J. Kyngäs: "Solving the person-based multitask shift generation problem with breaks", in Proc. of the 5th International Conference On Modeling, Simulation And Applied Optimization (ICMSAO), Hammamet, Tunis, 2013.

[22] S. Lin and B.W. Kernighan, "An effective heuristic for the traveling salesman problem", Operations Research 21, 1973, pp. 498–516.

[23] P. Rasmussen and M. Trick, "Round robin scheduling - A survey", European Journal of Operational Research 188, 2008, pp. 617-636.

[24] G. Kendall, S. Knust, C.C. Ribeiro and S. Urrutia, "Scheduling in Sports: An annotated bibliography", Computers and Operations Research 37, 2010, pp. 1-19.

[25] [8] D. de Werra, "Scheduling in sports", in Studies on graphs and discrete programming, edited by Amsterdam and Hansen, 1981, pp. 381-395.

[26] K. Easton, G. Nemhauser, and M. Trick "The traveling tournament problem: description and benchmarks" in Proc of the 7th. International Conference on Principles and Practice of Constraint Programming, Paphos, 2001, pp. 580–584.

[27] K. Nurmi, J. Kyngäs, and D. Goossens, "Scheduling the Highly Constrained Finnish Major Ice Hockey League" in Proc of the 11th Brazilian Congress on Computational Intelligence (CBIC), Recipe, Brazil, 2013. (submitted to review)

[28] K. Nurmi, D. Goossens, T. Bartsch, F. Bonomo, D. Briskorn, G. Duran, J. Kyngäs, J. Marenco, CC. Ribeiro, FCR. Spieksma, S. Urrutia and R. Wolf-Yadlin, "A Framework for Scheduling Professional Sports Leagues", in Ao, Sio-Iong (ed.): IAENG Transactions on Engineering Technologies Volume 5, Springer, USA, 2010.

[29] A.T. Ernst, H. Jiang, M. Krishnamoorthy, and D. Sier, "Staff scheduling and rostering: A review of applications, methods and models," European Journal of Operational Research 153 (1), 2004, pp. 3-27.

[30] N. Kyngäs, K. Nurmi and J. Kyngäs, "The Workforce Optimization Process Using the PEAST algorithm", Lecture Notes in Engineering and Computer Science: Proceedings of The International MultiConference of Engineers and Computer Scientists, Hong Kong, 2013.

[31] E.I. Ásgeirsson, J. Kyngäs, K. Nurmi and M. Stølevik, "A Framework for Implementation-Oriented Staff Scheduling", in Proc of the 5th Multidisciplinary Int. Scheduling Conf.: Theory and Applications (MISTA), Phoenix, USA, 2011.

Utilizing Bézier surface pattern recognition for analyzing military tactics

Jari Sormunen, Lt.Col.
National Defence University,
Finnish Defence Forces, Santahamina
Finland

Harri Eskelinen, D.Sc
Lappeenranta University of Technology
PL 20, 53850 Lappeenranta
Finland
harri.eskelinen@lut.fi

***Abstract*— This paper describes how to develop and utilize a method based on Bézier surface pattern recognition, which could be used for the overall military tactical analysis of a company's attack. This paper also explains how this method could be applied to an integrated analysis of the most important tactical factors affecting the success and task fulfillment of an attack together with the effects of the leader's decision-making and his or her tactical solutions. The presented approach aims towards overall optimizing process to ensure the performance maintenance of a military unit. The applied idea of Bézier surface pattern recognition gives options to widen the approach towards advanced computational analysis.**

Keywords*— *Military tactics; decision-making; company; attack exercise; Bézier surface

I. Introduction

The tactical basic research, "The Success Factors of Infantry Company's Attack" (SCA research), which was carried out in the Finnish Defence Forces (FDF) during the years 2004-2007 forms the empirical background of this paper. The SCA research focused on analyzing different individual effects of selected explanatory measured variables. The variables were selected from the areas of tactics, situational awareness, battle task load, human factors, background factors and response variables. During this research, 59 attacks by infantry companies were analyzed. This paper presents a comprehensive evaluation method for tactical analysis. The development of the method started from constructing the data collection matrix. Later on in this paper, the method is called the CMEP method (Command and control, Manoeuvre, Effect, Performance Maintenance) and its data collection matrix is entitled the CMEP matrix. This matrix is presented in Fig. 1. The construction and order of the rows and the columns of the CMEP matrix are based on the observations collected in the SCA research and supported by literature research [1], [3].

II. Qualitative Analysis

In the Finnish Defence Forces' Field Manual 2008 [2] the central tactical principles are described with the words, "consciously, actively, simply, concentratedly and continuously". The sequence of the tactical principles, which is the sequence of the rows in the CMEP matrix, is determined based on the causalities found from the literature and the qualitative analysis of the source data collected from the SCA research. According [2], tactical principles should always be integrated with the time dependent result. Based on these facts, the sequence of the rows in the CMEP matrix, should be "knowledge of adversary and own, activeness, simplicity, concentration of effect, reserve and exertion of it".

TACTICAL ELEMENTS/ PRINCIPLES	COMMAND AND CONTROL	MANOEUVRE	EFFECT	PERFORMANCE MAINTENANCE	Frequency Relative share Probability
KNOWLEDGE OF ADVERSARY AND OWN	**1 FREEDOM OF ACTION** Shaping the threat Possibilities of achieving effect Common tactical understanding Position and condition info Avoiding force binding	**2 TERRAIN AND CIRCUMSTANCES** Restrictions of adversary Own force strengths Dominant terrains Relative agility	**3 USE OF PERFORMANCE** Choosing the right object/target Avoiding passive formality Active use of deceptive actions Economic use of force Profitable synchronization	**4 CONCEALMENT** Variation of action styles Own battle plan secrecy Situation picture of adversary Effect-based decentralization	
ACTIVENESS	**5 SITUATION EVALUATION AND POSSIBILITIES** Seeing the possibility Courage to utilize Will to take initiative Reconnaissance pull Sensibility of thinking	**6 BATTLE READINESS** Adaptation to situations Responding to the unexpected Continuity of movement Leadability Positioning	**7 COMBAT TECHNIQUES** Firing situations (opening fire) Adversary on move when fired Own units positioned when firing Active use of battle composition Proper utilization of strengths	**8 FORCE PROTECTION** Keeping up preparedness Ensuring early warnings Own performance capacity Terrain adaptation Air protection and -defence	
SIMPLICITY	**9 CONCENTRATING ON THE ESSENTIAL** Main attention on adversary Battle task clear in mind Action competence (training) Utilization of own example Short communication (signals)	**10 EXECUTION CAPABILITY** Effect and speed (queues) Standardized starts (STP) Weapons quickly to positions Aim to concentrate effect Mobility improvement	**11 CLEAR INTENTION** Fulfilling the battle task Maximizing casualty production Minimizing own force casualties Satisfactory input-output ratio Battle task tactics	**12 FLEXIBILITY** Broad method spectrum Fast adaptability Speed of situation variation Security	
CONCEN-TRATION OF EFFECT	**13 BREAKING THE BATTLE PLAN OF ADVERSARY** Stopping the manoeuvre Forcing adversary to react Aiming at surprise Active choice of time and place	**14 BATTLE PULL** Advantageous positions Several effect directions Adjusting the movement Profitable battle formations Situational speed	**15 BATTLE SUPPORT** Producing enough casualties Overall profitable use of all fires Proper centre of gravity Co-effect of supporting branches Breaking adversary's cohesion	**16 BATTLE SERVICE** Material readiness Support to acting readiness Foreseeing needs Advanced fulfilment of needs Pushing principle in service	
RESERVE AND EXERTION OF IT	**17 PREPARING FOR VARIATION OF SITUATION** Readiness for the unexpected Active use of situation picture Evaluating options Searching culmination Effort to take initiative	**18 CONTINUITY** Keeping up initiative Aiming at battle culmination Attention to possibilities Focus after battle engagement Keeping up battle engagement	**19 UTILIZATION OF SUCCESS** Developing initiative Creating culmination Keeping up adversary's pressure Using the possibility windows Persistence	**20 RELEASING NEW RESERVE** Utilization of initiative Utilization of culmination Possibilities of continuity Developing battle situations Adversary's pressure growth	
Frequency Relative share Probability					

Fig. 1. The CMEP matrix.

According [2], the tactical elements are command and control, manoeuvre, effect and performance maintenance. The tactical principles become apparent in the interaction with the tactical elements of the battle. According to [2], the purpose of the manoeuvre near the adversary is to move one's own troops safely to the position from which it is possible to affect the adversary effectively with fire. Further, command and control can be seen as a factor which aims to integrate the other tactical elements of the battle field. When considering the chronological order of the tactical elements, the sequence is command and control, manoeuvre, effect and performance maintenance. The tactical elements (columns of the CMEP matrix) of the CMEP matrix are the same as the tactical elements in the battlefield according to [2].

The tactical elements presented in the columns of the CMEP matrix can be evaluated by utilizing the adverbs presented in the rows of the CMEP matrix (consciously, actively, simply, concentratedly and continuously). For example, command and control could simultaneously be conscious, active, simple, concentrated and/or continuous). This viewpoint has lead to the 5*4 - CMEP matrix and enables the qualitative tactical evaluation of an attack. From the tactical viewpoint, the central aspect which connects the rows and columns of the CMEP matrix is time. As illustrated in Fig. 2, the time dependent structure of the CMEP matrix makes it possible to study in the desired time window both what actions have been taken and in what way these actions were carried out. Within the CMEP matrix, a part of a tactical phenomenon could be expressed as "conscious command and control" in the time window "tw_1" or it could be entitled "active manoeuvre" in the time window "tw_2" if it is relevant to limit the size of the time window to illustrate just a small part of the tactical phenomenon. From the viewpoint of the battle result, the study of tactical phenomena could be extended in the time window "tw_3", which covers several checkers of the CMEP matrix (several tactical actions have taken place and are carried out in different ways). Because both the tactical elements and tactical principles are time dependent and there is causality which establishes the sequence of them, there is also a time axis passing along the diagonal of the CMEP matrix. The CMEP matrix construction makes it possible to describe the tactical grounds so that e.g. good command and control could be simultaneously based on situational awareness (consciousness) and activeness, or on the other hand, effect could be simultaneously concentrated and simple. In addition to the description of these types of separate tactical reason-consequence-relations, the CMEP matrix construction makes it possible to study comprehensively different integrated tactical phenomena within the 5*4 field (see Fig. 2). In this research the checkers, which are formed at the crossings of the rows and columns of the CMEP matrix, are called CTVs (Central Tactical Variable) and the aspects they can be divided into are called tactical items. The contents of the tactical variables and items are established by the qualitative analysis of the source data collected from the SCA research. The qualitative analysis is carried out by applying the principles presented in [4].

III. QUANTITATIVE ANALYSIS

The next logical step after the qualitative analysis, which resulted in the construction of the CMEP matrix, is the quantifying process of the collected data to be entered into the CMEP matrix as numerical values of each CTV. When quantifying the qualitative results, the numerical values of each CTV value were calculated by classifying the number of positive/ neutral/ negative perceptions. The source data comes from the SCA research [8]. The presentation can be developed in a more illustrative manner by constructing a surface model from these numerical values. This is performed by changing the numerical values of the matrix into the height values of the surface model. As an essential part of this quantitative analysis, formulated Bézier surfaces were used as resulting surfaces, which indicate the goodness of the tactical solutions and performance of an attack. In addition to the formulation of the surface, a specialized surface ratio curve is calculated to be able to evaluate the attacks by using quantified grading.

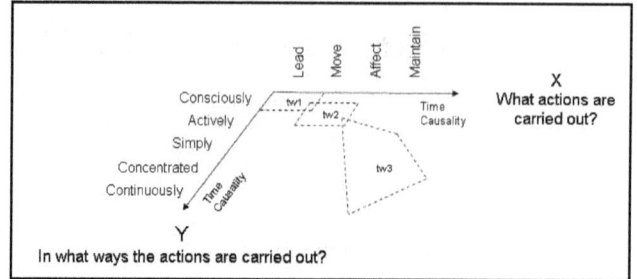

Fig. 2.   Utilization of the matrix construction for forming the tactical time windows ($tw_{1...3}$ = time window 1...3).

Based on the qualitative viewpoint, we already know how the columns of the CMEP matrix describe "what is done" (lead, move, affect, maintain) and how the rows of the CMEP matrix describe "in what ways these actions are carried out" (consciously, actively, simply, concentratedly, continuously) when a company carries out an attack. When we integrate the third dimension with the matrix (z-axis), it is possible to evaluate from the viewpoint of tactics the success of each action and the success of an attack. This principle is illustrated in Fig. 3. If we regard the numerical value of each central tactical variable as its assessment related to the success of the corresponding action described with this variable, the surface model constructed based on the numerical values of each CTV can be regarded as an overall assessment of the analyzed attack. The height values evaluate "how well" knowledge of the adversary and one's own force, activity, simplicity, concentration of effect and reserve and exertion of it have been carried out in relation with command and control, manoeuvre, effect and performance maintenance. The height values of the surface model change depending on time, which makes it possible to study the tactical phenomena at different moments during the progress of an attack or after the end of an attack. The quantitative evaluations of task fulfillment and casualty data were integrated into the presentation by positioning the surface at the z-axis according to the given numerical values of task fulfillment and casualty data. In the same way it would be possible e.g. to change the positioning of the surface depending on the difficulty level of the task. This quantifying process is presented in outline in Fig. 4.

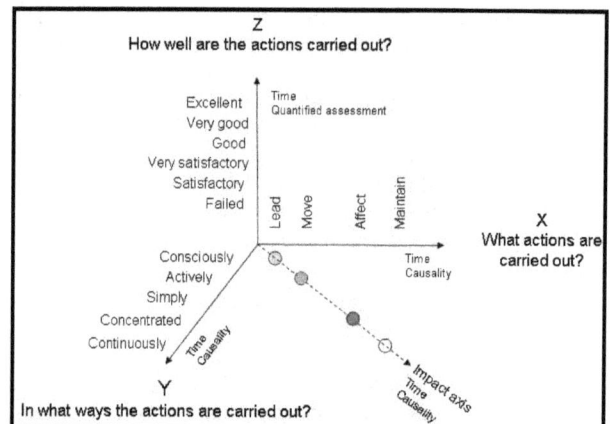

Fig. 3.   Integrating the z-axis with the CMEP matrix of the matrix. ($tw_{1...3}$ = time window 1...3).

As shown in Fig. 4, the basic height values of the surface are calculated by summing the scaled numerical values describing the task fulfilment and casualties. These values can be regarded as basic assessment criteria of success in the battle and therefore their scaled values are added to each central tactical variable to move the surface to the higher or lower level, which indicates how well the task is fulfilled and what the casualty ratio was after the battle. The additional quantified and scaled values of each central tactical variable are summed with these basic values.

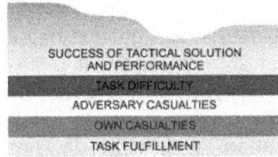

Fig. 4. Quantifying process

An example of the Bézier surface, which is calculated based on the measured data in the SCA research [8], for successful attacks is presented in Fig. 5.

IV. CALCULATING THE SURFACE AREA RATIO CURVE

From the produced surface model of the CMEP method, it is possible to cut different slices along xy-, yz- or zx-planes to evaluate different tactical aspects or to conduct a time dependent analysis of them. The calculated grading curve (surface area ratio curve) of the CMEP method makes it possible to carry out an exhaustive and overall evaluation of the attack exercise. To be able to evaluate the tactics of the battles by using quantified grading, the surface area ratio curve is calculated. An example is presented in Fig. 6. From this curve, it is reasonable to read how large a relative portion of the CTVs have reached the required grading level. On the other hand, it is possible to set a grading value which divides the battles into successful or unsuccessful ones. This curve also shows how wide the basis for success has been or if only a few CTVs affecting success have reached the required level. This grading curve is an application of the Abbott-Firestone curve, which is applied to surface analysis [5].

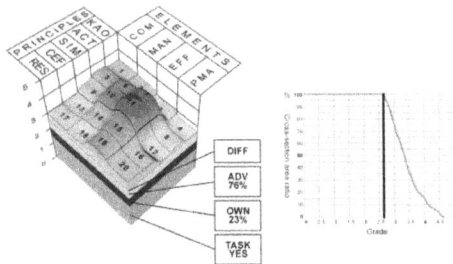

Fig. 5. Calculated Bézier surface and the grading curve.

V. BÉZIER SURFACE MODELING

Bézier surface presentations are utilized to support both decision-making and to integrate the results of several design or reasoning stages, e.g. in optimum shape design, according to [9]. Also in [6] the possibilities to utilize Bézier curves in

different types of practical applications have been evaluated to improve the local information gained from the surface model. In [6] improved and enhanced Bézier curve models are presented. Extensive research has also been conducted to find means to describe the shape information of Bézier surfaces [7] for interpreting different modeled phenomena.

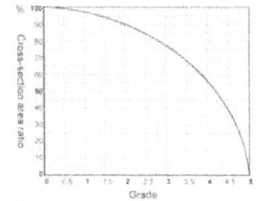

Fig. 6. The surface area ratio curve.

These aspects are of great interest when there is a need to integrate qualitative and quantitative measurement results to support tactical analysis. When evaluating the reliability aspects of the surface modeling, two main points were checked: How the selected Bézier surface calculation differed from other possible modeling techniques in relation to pattern recognition and what the difference is in grading. The viewpoint of interest focuses to the higher levels of the surface, which start above the layers describing one's own and the adversary's casualties together with task fulfillment. Four different surface modeling techniques were compared: Bézier, Mozaic, Pyramid and Block surfaces. Basically, the Mozaic model describes exactly the measured values of each item in all twenty CTVs. Compared to this, the Block model summarizes the values of different items within each of the twenty CTVs and it shows correctly the summarized values of each CTV. The Pyramid model is otherwise the same, but only the peaks are illustrated with the sharp vertex of twenty cones. The idea of the CMEP method is to combine time aspect into both CMEP matrix axes, which means that the heights of the neighboring bars should change smoothly according to the tactical phenomena. Unlike the other modeling techniques, the Bézier surface is able to illustrate this feature due to its mathematical properties. The comparison results of different grading curves and the surfaces of each modeling techniques are presented in Fig. 7.

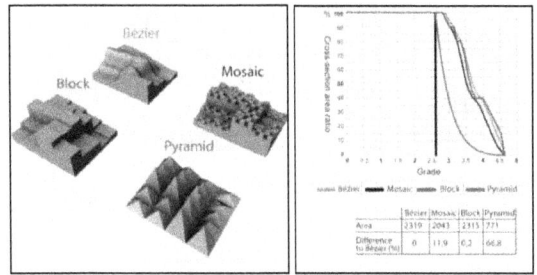

Fig. 7. Comparison of different surface modeling techniques.

VI. INTERPRETATION OF BÉZIER SURFACE PATTERNS

As a part of computing advances in military OR (Operations Research), especially in tactical decision-making, the surface interpretation is based on recognizing some geometrical shapes from the surface pattern. This recognizing process is supported by a picture series describing the most relevant surface shapes which are assumed to present some tactical phenomena and their success. The comparison of the calculated Bézier surface is performed with theoretical surfaces. Based on these theoretical surfaces, a set of definition maps were produced to support the interpretation process of military tactics. The comparison is divided into the following three parts: Geometric similarity of the definition maps and the calculated surface pattern, tactical dependency between researched viewpoints and the synthesis of the surface patterns. From the top view of the calculated surface, it can be proved that surface patterns at different height levels of the surface indicate and highlight by different weightings the aspects which have an effect on the result of the battle. It is as important to recognize these surface patterns at different height levels as it is to recognize the surface pattern at the highest (=most visible) surface level. This shows that several aspects simultaneously have an effect on the result of the battle. It also shows that depending on the battle situation, the impact factor of these aspects varies. As documented in Table 1, in this case example, only one chain is named and analyzed. Surface pattern recognition and interpretation. In this case, according to tactical interpretation, in order to enable initiative in battle, the leaders have ensured the freedom of action by maintaining the possibility to affect in creative ways and by avoiding force binding. In addition to this, the leaders have been open to looking for new possibilities and to taking initiative. Initiative in manoeuvres, effects and other actions has shown in proactive actions to varying situations and the unexpected actions of the adversary. The leaders have had a strong understanding of utilizing the unit's and its subunits' execution capability and they have understood the necessary functions to carry out the planned, prepared and trained actions before the adversary has had time to force them to do so. Both the leaders and the subunits have had the courage and the justification to find and use the effects of supporting branches in a simple and creative way to fulfill the battle task.

TABLE I.   SURFACE PATTERN RECOGNITION AND INTERPRETATION

Constructed surface and its top view and recognized pattern shape	Calculated Bézier	
	Top view	Schematic
Recognized shape from the top view of the surface	Recognized (visible) sequential CTVs	Nearest schematic geometry in the set of definition map
Chain	5 → 6 → 10 → 11 → 15 → 19	1 → 5 → 6 → 10 → 11 → 15

VII.   DISCUSSION AND CONCLUSIONS

The aim of the CMEP method is to integrate several different variables and affecting factors into one overall tactical analysis according to the leaders' decisions, solutions, and orders and according to the unit's actions and according to the events in the battle space. Using a Bezier surface for a data set is simple from mathematical point of view, and it can be easily done by using commercial software. On the other hand, visualization of data and objective functions is an essential part of decision-making. Therefore different interpolation methods help a decision-maker to understand the problem. By integrating a qualitative and quantitative analysis of an attack within the Bézier surface model, it is possible to extract and identify certain key points of the battle such as task fulfillment, required resources to achieve victory in a battle, tactical timing aspects, justification of tactical decisions and culmination points of the battle. Because the surfaces are produced from measured data into the form of Bézier surfaces, it is possible to add mathematical comparison and pattern recognition in the CMEP method. The primary challenge was in finding an appropriate way to handle several qualitative parameters describing tactical aspects. The sensitivity analysis of the CMEP method shows that it is possible to affect the resolution of the surface model by tuning the scaling of different layers of the surface model. However, a sufficient amount of source data is more critical to ensure that the height differences of the surface model are clear enough to illustrate different tactical aspects. This research will be continued within the cooperation between FDF and LUT. The next stage will be the further development of the interpretation maps of the Bézier surfaces..

ACKNOWLEDGMENT

Authors express their gratitude to Mr. Sami Karsikas for illustrative figures in this paper and to LUT Language Centre for proofreading.

REFERENCES

[1] C.Bellamy, The Evolution of Modern Land Warfare, Theory and Practice, Routledge, 13, pp. 15-21, 25, 1990.

[2] Field Manual, The Finnish Defence Forces, Infantry Company's Battle Guideline, Edita Prima Oy, pp. 18-20, 92-93 and 123-124, 2008.

[3] B.H Liddell Hart, Strategy - the indirect approach, New York, pp. 336-341, 1954.

[4] M.Miles and M.Huberman, Qualitative Data Analysis, 2nd Edition, Thousand Oaks, pp.10-12, 1994.

[5] J. Schmahling and F. Hamprecht, "Generalizing the Abbott-Firestone curve by two new surface descriptors", Wear, Vol. 262, Issue 11-12, pp.1360-1371, 2007.

[6] F.Sohel, G.Karmakar, L. Dooley and J.Arkinstall, "Enhanced Bezier curve models incorporating local information", IEEE International Conference on Acoustics, Speech, and Signal Processing, Vol. IV, article 1415993, pp. IV253-IV256, 2005.

[7] F.Sohel, G.Karmakar and L.Dooley, "Bezier curve-based character descriptor considering shape information", 6th IEEE/ACIS International Conference on Computer and Information Science, pp. 212-216, 2007.

[8] J.Sormunen and H.Eskelinen, Measured Tactics - Success Factors of Infantry Company's Attack, Edita Prima Oy, pp. 43-61, 2010.

[9] Vucina, D., Lozina, Z., and Pehnec, I. Computational procedure for optimum shape design based on chained Bézier surfaces parameterization, Engineering Applications of Artificial Intelligence, Vol 25, Issue 3, No. 4, 2012.

Sensitivity Analysis in Sequencing Models

Yury Nikulin

Department of Mathematics and Statistics

20014 University of Turku

Email: yurnik@utu.fi

Abstract—**A problem of minimization a liner form on a set of substitutions is considered. The main problem parameters are subject to small independent perturbations. Sensitivity analysis of an optimal solution is the main research focus. Appropriate measures of optimal solutions quality with respect to the uncertain environment are introduced. These measures correspond to the so-called stability and accuracy functions defined earlier in literature for some combinatorial optimization problems. Analytical expressions of such functions are presented. The maximum norms of perturbations for which an optimal solution preserves own optimality are also specified. The results are interpreted in terms of some well-known job sequencing models. The link between sensitivity analysis and robust optimization is established. Computational issues are discussed.**

I. INTRODUCTION

The increased interest (see [3]) in sensitivity and post-optimal analysis of various optimization and scheduling models is motivated by the fact that the input parameters, as a link between the real world and the model, are far from being unambiguously defined. Solutions of discrete optimization problems can exhibit remarkable sensitivity to perturbations in the space of parameters of the problem, thus often rendering a computed solution highly infeasible, suboptimal, or both. It is evident that an arbitrary real-life problem may not be correctly specified and solved without using the results of stability theory and related issues at least implicitly.

The terms stability and accuracy analysis, sensitivity analysis or post-optimal analysis are used for the stage at which a solution of an optimization problem has already been specified (the problem with initial data has been resolved to optimality), and additional calculations are performed in order to investigate how this solution depends on the numerical input data. In this context, sensitivity analysis may serve as an additional decision making tool helping to select one solution which has better stability (robustness) properties.

The most frequently considered object as a tool of sensitivity analysis is the so-called stability radius with respect to some given optimal solution. It gives a subset of problem parameters for which this solution remains optimal. There are already similar investigations in multiobjective case (see e.g. [2]). For example, one can find also a large survey on sensitivity analysis of vector unconstrained integer linear programming in [1].

It is important to note that the stability radius may not provide us with any information about the quality of a given solution in the case when problem data are outside of the stability region. Some attempts to study a quality of the problem solution in this case are connected with concepts of stability and accuracy functions. These functions were firstly introduced in [4] for scalar combinatorial optimization

problem. Later, in [5] the results of [4] were extended to the vector linear discrete optimization problem with Pareto and lexicographic optimality principles.

II. PROBLEM FORMULATION

First, we consider a single objective variant of the well-known problem of minimization a linear form on a set of substitutions. Let $m \geq 2$ be a problem dimension. Let $A = (a_1, a_2, \ldots, a_m) \in \mathbf{R}_+^m$ and $B = (b_1, b_2, \ldots, b_m) \in \mathbf{R}_+^m$ be two positive real-valued row vectors where $\mathbf{R}_+ = \{u \in \mathbf{R} : u > 0\}$. These vectors represent the problem parameters. Let S_m be a symmetric group of substitutions on a set $N_m = \{1, 2, \ldots, m\}$. The objective function is a linear function defined on a non-empty subset $T \subset S_m$:

$$f(t, A, B) = \sum_{j=1}^{m} a_j b_{t(j)},$$

where $t = (t(1), t(2), \ldots, t(m))$. We assume that the set T is known, fixed and can be specified by complete enumeration of the alternatives (substitutions). In practise, of course, the set of feasible substitutions can be specified by a set of constraints or some elimination (propagation) rules, however since in our model the feasibility is not going to be anyhow affected we assume full enumeration possibility as the simplest one. The problem, which we denote $Z(T, A, B)$, consists in finding $t^* = arg \min_{t \in T} f(t, A, B)$. We denote $T(A, B)$ the set of all optimal substitutions in problem $Z(T, A, B)$.

Many problems of scheduling theory such as e.g. job sequencing and routing problems as well as many classical combinatorial optimization problems can be represented as $Z(T, A, B)$ by making a relevant choice of basic problem parameters A, B and T. For example, consider a single machine scheduling problem where all jobs $1, 2, \ldots, m$ are sequenced to be processed on one machine (a unit disjunctive resource). Let A encode job processing times, and set T specify feasible (due to the possible presence of precedence constraints) orderings of the jobs. So, any particular $t = (t(1), t(2), \ldots, t(m)) \in T$ specifies for each job $j \in N_m$ the place $t(j)$ at which it has to be processed. Then given $B = (m, m - 1, \ldots, 1)$, for any particular $t \in T$, the value of $f(t, A, B)$ will be equal to the total sum of completion times $\sum_{j=1}^{m} C_j$. Indeed, since under t the job position in the sequence is given by $t(j)$ its processing time a_j will be counted (summed up) $m - j + 1$ time (due to additivity of total flow) which results in

$$f(t, A, B) = \sum_{j=1}^{m} (m - j + 1) a_{t^{-1}(j)} = \sum_{j=1}^{m} C_j,$$

Proceedings of FORS40 - 2013 Lappeenranta, Finland - ISBN 978-952-265-435-9

where t^{-1} is inverse to t, so $a_{t^{-1}(j)}$ is a processing time of the job placed at j-th position in the ordering.

For any fixed pair of substitutions t, $t' \in T$ we define

$$q(t, t', A, B) = f(t, A, B) - f(t', A, B) = \sum_{j=1}^{m} a_j (b_{t(j)} - b_{t'(j)}).$$

If t^* is optimal substitution in problem $Z(T, A, B)$, then $q(t^*, t', A, B) \leq 0$ for every $t' \in T$, and equality to zero happens in the presence of more than one optima. Denote $T(A, B)$ the set of all optima in $Z(T, A, B)$.

For a given substitution $t \in T$, denote

$$W(t, A, B) = \{t' \in T : \sum_{j=1}^{m} | b_{t(j)} - b_{t'(j)} | > 0\},$$

i.e. $W(t, A, B)$ represents the set of all substitutions $t' \in T$ which are different from t with respect to B in that sense that there exists at least one $j \in N_m$ with $b_{t(j)} \neq b_{t'(j)}$. If all b_j, $j \in N_m$ are different (as e.g. in the above mentioned case where $B = (m, m-1, \ldots, 1)$), then $W(t, A, B) = T \backslash \{t\}$. Any optimum $t^* \in T(A, B)$ with $W(t^*, A, B) = \emptyset$ we call trivial, otherwise nontrivial. The set of all nontrivial optima t^* denote as $\hat{T}(A, B)$. There are two cases possible only: either $\hat{T}(A, B) = T(A, B)$ (all optima are nontrivial) or $\hat{T}(A, B) = \emptyset$ (all optima are trivial).

When vector of coefficients of objective function changes, then initially optimal solution may become no longer optimal in the modified problem. We will evaluate the quality of this solution from the point of view of its robustness on data perturbations. Namely, we introduce for $t^* \in T(A, B)$ and a given vector $A \in \mathbf{R}_+^m$ so-called absolute error of this solution:

$$\varepsilon(t^*, A, B) := f(t^*, A, B) - \min_{t \in T} f(t, A, B) =$$

$$\max_{t \in T} q(t^*, t, A, B).$$

Observe that for arbitrary A and B, we have $\varepsilon(t^*, A, B) \geq 0$. If $\varepsilon(t^*, A, B) > 0$, then $t^* \notin T(A, B)$, and this positive value of the absolute error may be treated as a measure of inefficiency of t^* in $Z(T, A, B)$. It is also clear that for $t^* \in T(A, B)$

$$\varepsilon(t^*, A, B) = \max \left\{ 0, \max_{t \in W(t^*, A, B)} q(t^*, t, A, B) \right\} =$$

$$\max_{t \in \bar{W}(t^*, A, B)} q(t^*, t, A, B),$$

where $\bar{W}(t^*, A, B) := W(t^*, A, B) \cup \{t^*\}$. Just another implication of this fact is that while calculating the value of the absolute error it is sufficient to look at t with $\sum_{j=1}^{m} | b_{t^*(j)} - b_{t(j)} | > 0$ exclusively, since only they may contribute to positiveness of the error. Another simple observation is that if t^* is trivial, then so do the other optima (if any), and hence $\varepsilon(t^*, A, B) = 0$ for all $t^* \in T(A, B)$. Moreover this will be kept true for any modified $A \in \mathbf{R}_+^m$, so it is sufficient since now to perform sensitivity analysis for nontrivial optima only. Thus, after reasoning above, we may conclude that it could be a particular interest to measure the absolute error of nontrivial optimum $t^* \in \hat{T}(A, B)$ as specified above.

The following example will finalize this section.

Example 1: We illustrate the absolute error calculation using the problem of job sequencing on a single machine which has been already mentioned earlier in the text. Given $m = 3$ jobs which constitute the set $N_3 = \{1, 2, 3\}$, and let their processing times be encoded by vector $A = (1, 2, 5)$. As we know, in order to model the case of total completion time minimization we should set $B = (3, 2, 1)$. If no precedence constraints are specified, then the set of feasible solutions for our problem contains all the permutations of $\{1, 2, 3\}$, so $T = \{t_1, t_2, \ldots, t_6\}$, where $t_1 = (1, 2, 3)$, $t_2 = (1, 3, 2)$, $t_3 = (2, 1, 3)$, $t_4 = (2, 3, 1)$, $t_5 = (3, 1, 2)$, and $t_6 = (3, 2, 1)$.

Then, we calculate the objective values as follows $f(t_1, A, B) = 12$, $f(t_2, A, B) = 15$, $f(t_3, A, B) = 13$, $f(t_4, A, B) = 19$, $f(t_5, A, B) = 17$, $f(t_6, A, B) = 20$. Hence, the optimal substitution is t_1 which prescribes the jobs to be processed in optimal order according to the well-known shortest processing time first rule (see e.g. [8]). Then since $t_1 \in \hat{T}(A, b)$, we have $\varepsilon(t^*, A, B) = 0$.

Now consider the same problem with vector $\tilde{A} = (1 + \delta, 2, 5)$, where $\delta > 0$. Then $f(t_1, A, B) = 12 + 3\delta$, $f(t_2, A, B) = 15 + 3\delta$, $f(t_3, A, B) = 13 + 2\delta$, $f(t_4, A, B) = 19 + 2\delta$, $f(t_5, A, B) = 17 + \delta$, $f(t_6, A, B) = 20 + \delta$. So, now it becomes clear that until $\delta \leq 1$, optimality of t_1 is preserved, i.e. $t_1 \in T(\tilde{A}, B)$. If processing time of job 1 experience larger perturbations ($\delta > 1$), it will lose optimality and then the positive value of the absolute error $\varepsilon(t^*, \tilde{A}, B) > 0$ shows the speed of deviation from optimum, which is actually $t_3 \in T(\tilde{A}, B)$.

III. STABILITY AND ACCURACY FUNCTIONS

Henceforward, we assume that in any particular instance of $Z(T, A, B)$ the set T is fixed, vector B is also fixed, but vector A may vary or is estimated with errors. Moreover, it is assumed that the problem $Z(T, A, B)$ resolved to optimality, i.e. the sets $T(A, B)$ and $\hat{T}(A, B)$ are detected. In the following we are interested, in fact, in finding for any specified optimum $t^* \in \hat{T}(A, B)$ the maximum value of the error $\varepsilon(t^*, A', B)$ where A' belongs to some pre-specified set. Two particular cases are considered.

In the first case we are interested in absolute perturbations of the weights of elements and the quality of a given solution is described by the so-called *stability function*. For a given $\rho \geq 0$ the value of the stability function is equal to the maximum absolute error of a given solution under the assumption that no weights of elements are increased or decreased by more than ρ.

In the second case we deal with relative perturbations of weights. This leads to the concept of the *accuracy function*. The value of the accuracy function for a given $\delta \in [0, 1)$ is equal to the maximum absolute error of the solution t^* under the assumption that the weights of the elements are perturbed by no more than $\delta \cdot 100\%$ of their original values.

For a given $\rho \in [0, q(A))$, where $q(A) = \min\{a_j : j \in N_m\}$, we consider a set

$$\Omega(\rho, A) := \left\{ A' \in \mathbf{R}_+^m : | a'_j - a_j | \leq \rho, \, j \in N_m \right\}.$$

For an optimal solution $t^* \in \hat{T}(A, B)$ and $\rho \in [0, q(A))$, the value of the stability function is defined as follows:

$$S(t^*, \rho) := \max_{A' \in \Omega(\rho, A)} \varepsilon(t^*, A', B).$$

In a similar way, for a given $\delta \in [0, 1)$, we consider a set

$$\Theta(\delta, A) := \left\{ A' \in \mathbf{R}_+^m : \mid a'_j - a_j \mid \leq \delta a_j, \; j \in N_m \right\}.$$

For an optimal solution $t^* \in \hat{T}(A, B)$ and $\delta \in [0, 1)$, the value of the accuracy function is defined as follows:

$$A(t^*, \delta) := \max_{A' \in \Theta(\delta, A)} \varepsilon(t^*, A', B).$$

It is easy to check that $S(t^*, \rho) \geq 0$ for any $\rho \in [0, q(A))$ as well as $A(t^*, \delta) \geq 0$ for each $\delta \in [0, 1)$.

The following theorem gives analytical formulae for calculating stability and accuracy functions.

Theorem 1: The following statements are true.

(i) For an optimal solution $t^* \in \hat{T}(A, B)$ and $\rho \in [0, q(A))$, the stability function can be expressed by the formula:

$$S(t^*, \rho) = \max_{t \in \bar{W}(t^*, A, B)} \left\{ q(t^*, t, A, B) + \rho \sum_{j=1}^m \mid b_{t^*(j)} - b_{t(j)} \mid \right\}.$$

(ii) For an optimal solution $t^* \in \hat{T}(A, B)$ and $\delta \in [0, 1)$, the accuracy function can be expressed by the formula:

$$A(t^*, \delta) = \max_{t \in \bar{W}(t^*, A, B)} \left\{ q(t^*, t, A, B) + \right.$$

$$\left. \delta \sum_{j=1}^m a_j \mid b_{t^*(j)} - b_{t(j)} \mid \right\}.$$

Proof. We prove (i) first. Let $t^* \in \hat{T}(A, B)$. Using the definitions of stability function and absolute error as well as property of max operator, we have

$$S(t^*, \rho) = \max_{A' \in \Omega(\rho, A)} \varepsilon(t^*, A', B) =$$

$$\max_{A' \in \Omega(\rho, A)} \max_{t \in \bar{W}(t^*, A, B)} q(t^*, t, A', B) =$$

$$\max_{t \in \bar{W}(t^*, A, B)} \max_{A' \in \Omega(\rho, A)} q(t^*, t, A', B).$$

Once $t \in \bar{W}(t^*, A, B)$ is fixed, $\max_{A' \in \Omega(\rho, A)} q(t^*, t, A', B)$ is attained at vector $\tilde{A}$ with elements $\tilde{a}_j = a_j - \rho$ if $b_{t^*(j)} - b_{t(j)} < 0$, and $\tilde{a}_j = a_j + \rho$ if $b_{t^*(j)} - b_{t(j)} > 0$, and we set $\tilde{a}_j = a_j$ otherwise.

Then we continue

$$\max_{t \in \bar{W}(t^*, A, B)} \max_{A' \in \Omega(\rho, A)} q(t^*, t, A', B) =$$

$$\max_{t \in \bar{W}(t^*, A, B)} q(t^*, t, \tilde{A}, B) =$$

$$\max_{t \in \bar{W}(t^*, A, B)} \left\{ q(t^*, t, A, B) + \rho \sum_{j=1}^m \mid b_{t^*(j)} - b_{t(j)} \mid \right\}.$$

The last formally ends the proof of (i). Now we prove (ii). Indeed, repeating the same steps as above it will be sufficient to notice that once $t \in \bar{W}(t^*, A, B)$ is fixed,

$$\max_{A' \in \Theta(\delta, A)} q(t^*, t, A', B)$$

is attained at vector $\tilde{A}$ with elements $\tilde{a}_j = (1 - \delta)a_j$ if $b_{t^*(j)} - b_{t(j)} < 0$, and $\tilde{a}_j = (1 + \delta)a_j$ if $b_{t^*(j)} - b_{t(j)} > 0$, and we set $\tilde{a}_j = a_j$ otherwise. So, we get directly

$$A(t^*, \delta) = \max_{t \in \bar{W}(t^*, A, B)} \max_{A' \in \Theta(\delta, A)} q(t^*, t, A', B) =$$

$$\max_{t \in \bar{W}(t^*, A, B)} q(t^*, t, \tilde{A}, B) =$$

$$\max_{t \in \bar{W}(t^*, A, B)} \left\{ q(t^*, t, A, B) + \delta \sum_{j=1}^m a_j \mid b_{t^*(j)} - b_{t(j)} \mid \right\}.$$

This ends the proof.

Notice that analytical formulae specified in theorem 1 are based on enumerating all solutions $t \in \bar{W}(t^*, A, B)$, so in general they are hard to be computed. It is also clear that stability and accuracy functions are continuous piecewise linear functions of their arguments ρ and δ respectively.

Now consider the case when $t^* \in \hat{T}(A, B)$ is a nontrivial optimal substitution in the original problem $Z(T, A, B)$ implying $S(t^*, 0) = 0$ and $A(t^*, 0) = 0$. It is of special interest to know the extreme values of $\rho \in [0, q(A))$ and $\delta \in [0, 1)$ for which $S(t^*, \rho) = 0$ and $A(t^*, \delta) = 0$, respectively. These values determine maximum norms of perturbations which preserve the property of the given solution to be an optimum. They are close analogues of the so-called stability and accuracy radii introduced earlier for single and multiple objective combinatorial optimization problems (see e.g. [1]). Formally, the stability and accuracy radii $RS(t^*, A)$ and $RA(t^*, A)$ are defined in the following way:

$$RS(t^*, A) = \sup \left\{ \rho \in [0, q(A)) : \; S(t^*, \rho) = 0 \right\},$$

$$RA(t^*, A) = \sup \left\{ \delta \in [0, 1) : \; A(t^*, \delta) = 0 \right\}.$$

If these radii are equal to zero, then there exists arbitrary small perturbations of the original vector A such that the initial optimum t^* loses its optimality under such perturbations. Otherwise, the solution t^* remains optimality for any problem with vector $A' \in \Omega(\rho, A)$, $\rho < RS(t^*, A)$, or $A' \in \Theta(\delta, A)$, $\delta < RA(t^*, A)$.

It is clear, that if t^* is a trivial optimum, so its stability (accuracy) radius is always equal to the maximum admissible level of perturbations, i.e. $RS(t^*, A) = q(A)$, and $RA(t^*, A) = 1$. The next result, which is a straightforward consequence of theorem 1, specifies expressions for stability and accuracy radii of nontrivial optima.

Theorem 2: The following statements are true.

(i) For an optimal solution $t^* \in \hat{T}(A, B)$, its stability radius can be expressed by the formula:

$$RS(t^*, A) = \min \left\{ q(A), \min_{t \in W(t^*, A, B)} \frac{q(t, t^*, A, B)}{\sum_{j=1}^m \mid b_{t^*(j)} - b_{t(j)} \mid} \right\}.$$

(ii) For an optimal solution $t^* \in \hat{T}(A, B)$, its accuracy radius can be expressed by the formula:

$$RA(t^*, A) = \min\left\{1, \min_{t \in W(t^*, A, B)} \frac{q(t, t^*, A, B)}{\sum_{j=1}^{m} a_j \mid b_{t^*(j)} - b_{t(j)} \mid}\right\}.$$

We illustrate calculating stability and accuracy functions as well as corresponding radii using the following example.

Example 2: Let the primitives be the same as in the previous example. Since $t_1 \in \hat{T}(A, b)$ is the only nontrivial optimum in the problem, we calculate $W(t^*, A, B) = \{t_2, \ldots, t_6\}$, $\bar{W}(t^*, A, B) = T = \{t_1, t_2, \ldots, t_6\}$, $\rho \in [0, 1)$, $\delta \in [0, 1)$, and hence corresponding values computed according to analytical formulae specified by theorems 1 and 2 are the following:

$$S(t^*, \rho) = \max\{0, -1 + 2\rho\}; \ RS(t^*, A) = 0.5;$$

$$A(t^*, \delta) = \max\{0, -3 + 7\delta, \ -1 + 3\delta, \ -7 + 13\delta,$$

$$-5 + 9\delta, \ -8 + 12\delta\}; \ RA(t^*, A) = \frac{1}{3}.$$

The meaning of numerical values of $RS(t^*, A)$ and $RA(t^*, A)$ will become even more transparent if we recall that the problem is a single machine job processing problem where optimal job processing sequence should obey shortest processing time first rule. In these terms, $RS(t^*, A)$, $t^* = (1, 2, 3)$, should be equal to a maximum value of $\rho \in [0, 1)$ such that $(1 \pm \rho, 2 \pm \rho, 5 \pm \rho)$ keeps the ordering that the first job has the smallest processing time, the second job has the second smallest processing time, and the third job has the largest processing time. Obviously, such a value is $\rho = 0.5$ Analogously, $RA(t^*, A)$, $t^* = (1, 2, 3)$, should be equal to a maximum value of $\delta \in [0, 1)$ such that $(1 \pm 1\delta, 2 \pm 2\delta, 5 \pm 5\delta)$ keeps the ordering that the first job has the smallest processing time, the second job has the second smallest processing time, and the third job has the largest processing time. Obviously, such a value is $\delta = \frac{1}{3}$.

Notice also that in [9], [10] an uncertain single machine scheduling problem with a criterion of minimization the weighted flow time is considered and attacked using some alternative tools of sensitivity analysis, such as dominance and stability box.

IV. CONCLUSION

The results presented in previous sections suggest that small changes or inaccuracies in estimating basic problem data may have significant influence on the set of optima preserving or destroying their optimality. The simplest measure of the solution's robustness is its stability and accuracy radius. But frequently having just a value of radius is not sufficient to rank the optima. Therefore, it is necessary to calculate complementary more general characteristics of solutions like stability and accuracy functions (see e.g. [7]). The other big challenge in robust and sensitivity analysis is to construct efficient algorithms to calculate the analytical expressions. To the best of our knowledge there are not so many results (see e.g. [6]) known in that area, and moreover some of those results which have been already known, put more questions than answers. It seems that calculating exact values is an extremely difficult task in general, so one could concentrate on either finding "easy" computable classes of problems or developing general metaheuristic approaches. Indeed, practical verification of stability conditions, calculating quantitative measures such as stability and accuracy radii (or functions) as well as finding robust solutions in general case can be at least as hard as to solve the problem itself (or even much harder). Nevertheless more methodological results might be developed and implemented for special cases of combinatorial optimization and scheduling problems with restrictions to some factors, such as structure of initial data, perturbations of particular problem parameters etc. As possible continuation of the research within this topic, it would be interesting to explore all these possibilities.

Acknowledgements

The results presented in this work were obtained during author's visit to the Granö-Centre at Tartu, Estonia. The visit became possible thank to kindly provided support from the Turku University Foundation.

REFERENCES

[1] V. Emelichev, E. Girlich, Y. Nikulin, D. Podkopaev, (2002). "Stability and regularization of vector problems of integer linear programming". *Optimization 51*, 645 – 676.

[2] V. Emelichev, D. Podkopaev, (2010). "Quantitative stability analysis for vector problems of 0-1 programming". *Discrete Optimization 7*, 48 – 63.

[3] H. Greenberg, (1998). "An annotated bibliography for post-solution analysis in mixed integer and combinatorial optimization." In D. Woodruff (ed.), Advances in computational and stochastic optimization, Logic programming and heuristic search, 97 – 148.

[4] M. Libura, (1999). "On accuracy of solution for combinatorial optimization problems with perturbed coefficients of the objective function". *Annals of Operation Research 86*, 53 – 62.

[5] M. Libura, Y. Nikulin, (2006). "Stability and accuracy functions in multicriteria linear combinatorial optimization problems". *Annals of Operations Research 147*, 255 – 267.

[6] M. Libura, E.S. van der Poort, G. Sierksma, J.A.A van der Veen, (1998). "Stability aspects of the traveling salesman problem based on k-best solutions", *Discrete Applied Mathematics 87*, 159 – 185.

[7] Y. Nikulin, O. Karelkina, M.M. Mäkelä, (2013). On accuracy, robustness and tolerances in vector Boolean optimization. *European Journal of Operational Research 224*, 449 – 457.

[8] M. Pinedo, *Scheduling: Theory, Algorithms, and Systems*, Prentice-Hall, Englewood Cliffs, NJ, USA, 1996.

[9] Y.N. Sotskov, T.-C. Lai, (2012) "Minimizing total weighted flow time under uncertainty using dominance and a stability box". *Computers and Operations Research 39*, 1271 – 1289.

[10] Y.N. Sotskov, N.G. Egorova, T.-C. Lai, (2009). "Minimizing total weighted flow time of a set of jobs with interval processing times", *Mathematical and Computer Modelling 50*, 556 – 573.

Equilibrium Based Flow Control in Wireless Networks with Moving Users

Igor V. Konnov
Department of System Analysis and
Information Technologies,
Kazan Federal University,
Kazan, 420008 Russia
Email: konn-igor@yandex.ru

Erkki Laitinen
Department of Mathematical Sciences
University of Oulu
90014 Oulu, Finland
Email: erkki.laitinen@oulu.fi

Olga Pinyagina
Department of Data Analysis
and Operations Research
Kazan Federal University
Kazan 420008, Russia
Email: Olga.Piniaguina@ksu.ru

Abstract—We first consider a general problem of information flows distribution calculation in a communication network with moving nodes. By using approximations of frequency values for nodes we formulate the above problem as a non-stationary variational inequality which also reduces to an optimization one. By solving each approximate problem within some tolerances we create a sequence of flows tending to a solution to the basic stationary problem.

Index Terms—Communication wireless networks, moving nodes, flows distribution, equilibrium approach, non-stationary optimization, iterative methods.

I. INTRODUCTION

The class of network flow control problems in communication networks is rather large and thoroughly investigated; see, e.g., [1]. Recently, the development of wireless telecommunication networks which provide faster availability and better information quality, support a wide range of new services. However, their behavior is rather complex, they have new specific features such as the absence of predefined physical links between nodes, mobility of nodes, and boundedness of the batteries capacity caused a lot of new problems; see, e.g., [2]. Because of the stochastic character of nodes movement and absence of any central decision maker, the most solution methods for such problems are mostly heuristic, hence their substantiation is deduced from simulation procedures. However, we believe that optimization and equilibrium approaches based on the corresponding mathematical models may be useful for creation of more efficient network control decisions.

In this paper, we treat a problem of flows distribution in a communication network with moving nodes as a non-stationary variational inequality via a suitable equilibrium approach. That is, we calculate (on-line) frequency values for nodes to be at cells constituting the network region as approximations of the corresponding probabilities. In turn, this variational inequality also reduces to an optimization one. This approach can be viewed as further development of that from [3], where each moving (user) node was treated

In this work, the first author was supported by grant No. 259488 from Academy of Finland and by the RFBR grant, project No. 13-01-00029a. The second author was supported by grant No. 259488 from Academy of Finland. The third author was supported by the RFBR grant, project No. 13-01-00029a.

as an independent Markovian chain. In fact, it enables us to cover much more general classes of applications. Next, by solving each approximate optimization problem within some tolerances we create a sequence of flows tending to a solution to the basic problem of flows distribution.

II. STATIC NETWORK EQUILIBRIUM MODEL

The model is determined on a communication network involving a finite set of nodes (users) $\mathcal{N}$, who are distributed within a region $\mathcal{R}$, which is supposed to be rectangular for simplicity. That is, $\mathcal{R}$ is represented by an $m \times n$ matrix, whose elements are equal square cells c_{ij} for $i = 1, \ldots, m$ and $j = 1, \ldots, n$. Next, we select a subset of origin-destination (O/D) pairs $\mathcal{W}$ among all the pairs of users, i.e. $\mathcal{W} \subseteq \mathcal{N} \times \mathcal{N}$.

First we describe the model for the static case where the network has some pre-defined and fixed topology given by a set of communication links (arcs) $\mathcal{A}$ which join the users (nodes); see, e.g., [1], [4], [5]. This means that each user has fixed position in $\mathcal{R}$. Hence, for each pair $w \in \mathcal{W}$ one can define the set of paths $\mathcal{P}_w$ joining this pair via the path-arc incidence matrix A with the elements

$$\alpha_{pa} = \begin{cases} 1 & \text{if arc } a \text{ belongs to path } p, \\ 0 & \text{otherwise.} \end{cases}$$

Besides, each (O/D) pair $w \in \mathcal{W}$ is associated with a non-negative flow demand b_w. Denote by x_p the path flow for path p. Then, the feasible set of network path flows is defined as follows:

$$X = \left\{ x \ \middle| \ \sum_{p \in \mathcal{P}_w} x_p = b_w, x_p \geq 0, \ p \in \mathcal{P}_w, w \in \mathcal{W} \right\}. \quad (1)$$

The set X has very simple structure. Namely, it is the Cartesian product of the simplices. We should now define a cost function. Given flow values x_p, one can determine the value of the arc flow

$$f_a = \sum_{w \in \mathcal{W}} \sum_{p \in \mathcal{P}_w} \alpha_{pa} x_p \quad (2)$$

for each arc $a \in \mathcal{A}$. We assume that for each link a we know a continuous function T_a, whose value $T_a(f_a)$ represents the delay in traversing link a with one unit of flow when the flow value on this arc is f_a or the marginal delay for this arc in the communication network. The summary cost at path flows vector x for path p has the form:

$$G_p(x) = \sum_{a \in \mathcal{A}} \alpha_{pa} T_a(f_a).$$

The equilibrium condition for such a network is formulated as follows. The feasible flow vector $x^* \in X$ is said to be an equilibrium vector if it satisfies the following conditions:

$$\forall w \in \mathcal{W}, \ q \in \mathcal{P}_w, \ x_q^* > 0 \Longrightarrow G_q(x^*) = \min_{p \in \mathcal{P}_w} G_p(x^*).$$

This formulation is based on the user-optimization principle, which asserts that a network equilibrium is established when no OD pair may decrease his/her traveling cost by making the unilateral decision to change his/her routes. However, it is equivalent to the variational inequality problem (VI for short): Find $x^* \in X$ such that

$$\begin{aligned} &\langle G(x^*), x - x^* \rangle \\ &= \sum_{w \in \mathcal{W}} \sum_{p \in \mathcal{P}_w} G_p(x^*)(x_p - x_p^*) \geq 0 \quad \forall x \in X. \end{aligned} \tag{3}$$

Moreover, due to the diagonality and continuity of the mappings T_a, they must be integrable, i.e. there exist the functions

$$\mu_a(f_a) = \int_0^{f_a} T_a(\tau) d\tau$$

for $a \in \mathcal{A}$. Then we can replace VI (3) by the optimization problem:

$$\min_{x \in X} \rightarrow \sum_{w \in \mathcal{W}} \sum_{p \in \mathcal{P}_w} \sum_{a \in \mathcal{A}} \alpha_{pa} \mu_a(f_a), \tag{4}$$

where f_a is defined in (2). That is, each solution of (4) solves (3), but the reverse assertion is true if each T_a is monotone.

There exist many iterative methods for solving VI (3) or problem (4), they are mostly closely related with gradient projection or conditional gradient ones; see, e.g., [6], [1], [4], [5]. They enables one to find a solution within any prescribed accuracy $\varepsilon > 0$ in a finite number of iterations. As a stopping criterion, one can use the condition that the distance between the neighbor iteration points does not exceed a given value ε.

III. DYNAMIC NETWORK EQUILIBRIUM MODEL

Now we turn to the general dynamic case where users can change their positions within the region $\mathcal{R}$. Note that networks with complex and non-stationary behavior of users (nodes) are typical for modern wireless communication systems; see e.g. [2]. Then the above classical formulation of the network equilibrium problem is not applicable because we do not know the topology of the communication network, in particular, the sets of arcs $\mathcal{A}$ and paths $\mathcal{P}_w$ and the path-arc incidence matrix A. Moreover, the delay value $T_a(f_a)$ becomes also unknown.

In order to overcome these difficulties, each moving (user) node was suggested to be treated as an independent Markovian chain [3]. That is, the basic time period was divided into equal slots $t = 1, 2, \ldots$, so that knowing the initial probabilities for each user to be in cells and transition probabilities one

can calculate probabilities for each user to be in each cell for any time slot and stationary probabilities. By introducing a suitable audibility threshold value one can calculate the topology data and delay function values and utilize them in a modified network equilibrium model; see also [7], [8]. However, this approach has certain restrictions because only Markovian chains without periodic classes can have stationary probabilities; see, e.g., [9].

Now we intend to extend the previous approach to more general classes of applications. To this end, we propose to calculate (on-line) frequency values for nodes to be in cells constituting the network region as approximations of the corresponding probabilities.

Thus, we suppose that all the users can move within the region $\mathcal{R}$, but only one transition (cell change) is possible during one time slot. The current state of user i at time slot t is described by the matrix

$$P^{(i,t)} = \left(\pi_{kl}^{(i,t)} \right) \tag{5}$$

where $\pi_{k,l}^{(i,t)}$ denotes the current frequency probability for the i-th user to be in cell c_{kl} at time slot t for $k = 1, \ldots, m$ and $l = 1, \ldots, n$. Hence, if the i-th user was $s(t)$ times in cell c_{kl} for time slots $k = 1, 2, \ldots, t$, then $\pi_{k,l}^{(i,t)} = s(t)/t$. Clearly,

$$\sum_{k=1}^m \sum_{l=1}^n \pi_{kl}^{(i,t)} = 1$$

for any i and t. The initial matrix $P^{(i,1)}$ can be defined as follows

$$\pi_{kl}^{(i,1)} = \begin{cases} 1 & \text{if the } i\text{-th user is in cell } c_{kl}, \\ 0 & \text{otherwise;} \end{cases}$$

i.e., elements of the matrix are defined by the initial position of each user.

(Assumption 1) In what follows we assume that the sequence of matrices $\left\{ P^{(i,t)} \right\}$ tends to some stable (limit) probability distribution matrix

$$\bar{P}^{(i)} = \left(\bar{\pi}_{kl}^{(i)} \right)$$

as $t \to \infty$ for each $i \in \mathcal{N}$. Clearly, this assumption is weaker essentially than those in [3].

We suggest to take the (limit) stationary probabilities for creating the network equilibrium problem, which is still based on model (1)–(3) similarly to the Markovian chain approach. So, given the matrix $\bar{P}^{(i)}$, we introduce for simplicity a distance threshold γ and then for each pair of nodes i, j we calculate the mean distance

$$u_{ij} = \sum_{k=1}^m \sum_{l=1}^n \sum_{\kappa=1}^m \sum_{\lambda=1}^n \bar{\pi}_{kl}^{(i)} \bar{\pi}_{\kappa\lambda}^{(j)} \rho((k,l),(\kappa,\lambda)),$$

where $\rho((k,l),(\kappa,\lambda))$ denotes the distance between the centers of cells (k,l) and (κ,λ). Hence, if $u_{ij} \leq \gamma$, then arc $a = (i,j)$ is included in the set of communication links $\bar{\mathcal{A}}$.

In such a way we determine the path-arc incidence matrix $\bar{A}$ with the elements

$$\bar{\alpha}_{pa} = \begin{cases} 1 & \text{if arc } a \text{ belongs to path } p, \\ 0 & \text{otherwise;} \end{cases}$$

and, for any selected (O/D) pair $w \in \mathcal{W}$ we can define the set of paths $\bar{\mathcal{P}}_w$ joining this pair. As above, we also denote by b_w non-negative flow demand for this pair and by x_p the path flow for path p. Then, the feasible set of network path flows is defined as follows:

$$\bar{X} = \left\{ x \mid \sum_{p \in \bar{\mathcal{P}}_w} x_p = b_w, x_p \geq 0, \ p \in \bar{\mathcal{P}}_w, w \in \mathcal{W} \right\}. \quad (6)$$

Next, we determine the value of the arc flow

$$f_a = \sum_{w \in \mathcal{W}} \sum_{p \in \bar{\mathcal{P}}_w} \bar{\alpha}_{pa} x_p \quad (7)$$

for each arc $a \in \bar{A}$. We assume that for each link $a = (i,j)$ and for each pair of cells (k,l) and (κ, λ) we know a continuous function $T_a^{\{(k,l),(\kappa,\lambda)\}}$, whose value $T_a^{\{(k,l),(\kappa,\lambda)\}}(f_a)$ gives the delay in traversing link a with one unit of flow when user i is situated in cell (k,l), user j is situated in cell (κ, λ), and the flow value on this arc is f_a. The summary cost at path flows vector x for path p has the form:

$$\bar{G}_p(x) = \sum_{k=1}^{m} \sum_{l=1}^{n} \sum_{\kappa=1}^{m} \sum_{\lambda=1}^{n}$$

$$\left(\sum_{a=(i,j)\in\bar{A}} \bar{\pi}_{kl}^{(i)} \bar{\pi}_{\kappa\lambda}^{(j)} \bar{\alpha}_{pa} T_a^{\{(k,l),(\kappa,\lambda)\}}(f_a) \right).$$

The equilibrium condition for the network is formulated as above. Hence, we can replace it by the following VI: Find $x^* \in \bar{X}$ such that

$$\langle \bar{G}(x^*), x - x^* \rangle \quad (8)$$

$$= \sum_{w \in \mathcal{W}} \sum_{p \in \bar{\mathcal{P}}_w} \bar{G}_p(x^*)(x_p - x_p^*) \geq 0 \quad \forall x \in \bar{X};$$

cf. (1)–(3). Besides, by using the integrability of T_a, we can also associate (6)–(8) with a suitable optimization problem as in (4).

However, the limit matrix $\bar{P}^{(i)}$ remains unknown for us and we can not solve problem (6)–(8) directly. Nevertheless, we can utilize its approximation at time slot t by using the matrix $P^{(i,t)}$ in (5) for $t = 1, 2, \dots$

In fact, for each pair of nodes i, j we can now calculate the mean distance

$$u_{ij}^{(t)} = \sum_{k=1}^{m} \sum_{l=1}^{n} \sum_{\kappa=1}^{m} \sum_{\lambda=1}^{n} \pi_{kl}^{(i,t)} \pi_{\kappa\lambda}^{(j,t)} \rho((k,l),(\kappa,\lambda)).$$

If $u_{ij}^{(t)} \leq \gamma$ for the distance threshold γ, then arc $a = (i,j)$ is included in the set of communication links $\mathcal{A}^{(t)}$ which approximate the set $\bar{A}$ for time slot t. After we determine the path-arc incidence matrix $A^{(t)}$ with the elements

$$\alpha_{pa}^{(t)} = \begin{cases} 1 & \text{if arc } a \text{ belongs to path } p, \\ 0 & \text{otherwise;} \end{cases}$$

and, for any selected (O/D) pair $w \in \mathcal{W}$ we can define the set of paths $\mathcal{P}_w^{(t)}$ joining this pair. As above, we denote by b_w non-negative flow demand for this pair and by $x_p^{(t)}$ the path flow for path p. Then, we define the t-th approximation of the feasible set

$$X^{(t)} = \left\{ x^{(t)} \mid \begin{matrix} \sum_{p \in \mathcal{P}_w^{(t)}} x_p^{(t)} = b_w, \\ x_p^{(t)} \geq 0, \ p \in \mathcal{P}_w^{(t)}, w \in \mathcal{W} \end{matrix} \right\}. \quad (9)$$

and value of the arc flow

$$f_a^{(t)} = \sum_{w \in \mathcal{W}} \sum_{p \in \mathcal{P}_w^{(t)}} \alpha_{pa}^{(t)} x_p^{(t)} \quad (10)$$

for each arc $a \in \mathcal{A}^{(t)}$. We also assume that for each link $a = (i,j)$ and for each pair of cells (k,l) and (κ, λ) there exists a continuous delay function $T_a^{\{(k,l),(\kappa,\lambda)\}}$ and define the summary cost at path flows vector $x^{(t)}$ for path p

$$G_p^{(t)}(x^{(t)}) = \sum_{k=1}^{m} \sum_{l=1}^{n} \sum_{\kappa=1}^{m} \sum_{\lambda=1}^{n}$$

$$\left(\sum_{a=(i,j)\in\mathcal{A}^{(t)}} \pi_{kl}^{(i,t)} \pi_{\kappa\lambda}^{(j,t)} \alpha_{pa}^{(t)} T_a^{\{(k,l),(\kappa,\lambda)\}}(f_a^{(t)}) \right).$$

The network equilibrium problem is then written in the form of the following VI: Find $x^{(t,*)} \in X^{(t)}$ such that

$$\langle G^{(t)}(x^{(t,*)}), x^{(t)} - x^{(t,*)} \rangle$$

$$= \sum_{w \in \mathcal{W}} \sum_{p \in \mathcal{P}_w^{(t)}} G_p^{(t)}(x^*)(x_p^{(t)} - x_p^{(t,*)}) \geq 0 \quad (11)$$

$$\forall x^{(t)} \in X^{(t)};$$

it represents the t-th approximation of the limit VI (6)–(8).

Clearly, the dynamic network can change its topology over time, but we add only one additional connectivity assumption, which indicates that the threshold γ is chosen properly.

(**Assumption 2**) For each (O/D) pair $w \in \mathcal{W}$ there exists at least one simple chain of directed links from $\mathcal{A}^{(t)}$ connecting the origin and destination nodes for each $t = 1, 2, \dots$

Thus, instead of the basic limit VI (6)–(8) we have a sequence of its approximations of form (9)–(11) for $t = 1, 2, \dots$ Such problems are called non-stationary; see, e.g., [10, Ch. VI]. Recently, new convergence results for inexact solutions of approximate problems to a solution of the limit non-stationary problem were established under rather mild conditions in [11], [12].

IV. SOLUTION METHOD

Being based on the previous results, we suggest a two-level method for the basic VI (6)–(8), which allows us to successively obtain approximations of equilibrium distribution of network flows, consecutively collecting and using the information about network users behavior changes. Each iteration of the method corresponds to some time slot t, whereas a lower level method solves the t-th approximate VI (9)–(11) within some tolerances.

Upper Level Method.

Step 0. Let some number $\delta > 0$ and a sequence of positive numbers $\{\varepsilon_t\}$ (tolerances), and probability matrices $P^{(1,i)}$ for each $i \in \mathcal{N}$ be given. We set $t = 1$ and choose an initial point $y^{(1)} \in X^{(1)}$.

Step 1. Using the point $y^{(t)}$ we solve VI (11) by Algorithm P described below within the tolerance ε_t and obtain an approximate solution denoted by $x^{(t,\varepsilon_t)}$.

Step 2. We make the transition to the next state of the network. We find new positions of users and calculate the matrices $P^{(i,t+1)}$ for all $i \in \mathcal{N}$.

Step 3. If $\|P^{(i,t)} - P^{(i,t+1)}\| < \delta$ for all $i \in \mathcal{N}$, then we obtain the desired accuracy of calculations and stop the iterative process. Otherwise we set $y^{(t+1)} = x^{(t,\varepsilon_t)}$, $t = t + 1$ and go to Step 1.

We suggest the following version of the gradient projection method as Algorithm P for Step 1 of the upper level method. For brevity, we denote by $\pi_{(t)}$ the projection operator onto the set $X^{(t)}$.

Algorithm P.

Step 0. Let some tolerance $\varepsilon_t > 0$, parameters $\alpha \in (0,1)$, $\beta \in (0,1)$, and an initial point $z^0 \in X^{(t)}$ be given. We set $k = 0$.

Step 1. If $\|z^k - \pi_{(t)}[z^k - G^{(t)}(z^k)]\| = 0$, then we obtain the solution to problem (11) and stop the process.

Step 2. We find the smallest nonnegative integer m which satisfies the inequality

$$\langle G^{(t)}(z^{k,m}), z^{k,m} - z^k \rangle \leq \alpha \langle G^{(t)}(z^k), z^{k,m} - z^k \rangle,$$

where $z^{k,m} = \pi_{(t)}[z^k - \beta^m G^{(t)}(z^k)]$.

Step 3. We set $z^{k+1} = z^{k,m}$. If $\|z^{k+1} - z^k\| \leq \varepsilon_t$, we obtain the desired accuracy and stop the iterative process. Otherwise we set $k = k + 1$ and go to Step 1.

We observe that the initial point z^0 in Algorithm P is taken from $y^{(t)}$ by some adjustment, say, as its projection onto $X^{(t)}$.

V. COMPUTATIONAL EXPERIMENTS

In order to verify the method proposed we performed preliminary numerical experiments on test problems, varying parameters related to dimensionality and accuracy. For the sake of simplicity we took the parameter γ large enough so that each pair of nodes was connected by two opposite-directed links, i.e. we have full graphs. Also, we took simple neighbor walking rules for moving users. The parameters of Algorithm P were taken as follows: $\alpha = 0.5$, $\beta = 0.5$, besides we chose $\varepsilon_t = \max\{1/t, 10^{-4}\}$. We report the total number of iterations and calculation time. The results of computations for different data are given in Table V. The numerical experiments were conducted using Microsoft Visual Studio 2005 (C++) and the hardware platform AMD Athlon 64 X2 Dual, 2.3 GHz, 2 Gb. In general, the results show rather satisfactory convergence.

δ	$N = 20$	$N = 40$
0.01	88 it., <1 s	87 it., 3 s
0.001	866 it., 2 s	866 it., 30 s
0.0001	8660 it., 20 s	8660 it., 322 s

TABLE I
TEST PARAMETERS: N IS THE TOTAL NUMBER OF NODES, δ IS THE
CHOSEN ACCURACY.

VI. CONCLUSIONS

We considered a problem of flows distribution in a communication network with moving nodes as a non-stationary variational inequality via a suitable equilibrium approach. This approach enables us to cover very general classes of applications. Inexact solutions of each approximate problem yields a sequence of flows tending to a solution to the basic problem of flows distribution. We suggested a two-level method for the basic problem, which allows us to successively obtain approximations of this flows distribution. The numerical experiments showed the applicability of the proposed approach to wireless networks with dynamic structure.

REFERENCES

[1] D. Bertsekas and R. Gallager, *Data Networks*, Prentice Hall, Englewood Cliffs, 1987.

[2] X. Cheng, X. Huang, and D.-Z. Du, Eds., *Ad Hoc Wireless Networking*, Kluwer, Dordrecht, 2004.

[3] I.V. Konnov and O. A. Kashina, "Optimization based flow control in communication networks with moving nodes", *Proc. of The Fourth Moscow Conf. on Operat. Res.*, MaxPress, Moscow, 2004, pp.116–118.

[4] A. Nagurney, *Network Economics: A Variational Inequality Approach*, Kluwer, Dordrecht, 1999.

[5] I.V. Konnov, *Equilibrium Models and Variational Inequalities*, Elsevier, Amsterdam, 2007.

[6] D. Bertsekas, "Projection methods for variational inequalities with application to the traffic assignment problem", *Math. Progr. Study*, 1982, vol.17, pp.139–159.

[7] I.V. Konnov, O.A. Kashina, and K. Lieska, "Optimization based flow control in communication networks", in *Proc. of Europ. Congr. on Comput. Meth. and Eng.*, ECCOMAS 2004, Jyvaskyla, 2004, pp.1–9 (http://www.mit.jyu.fi/eccomas2004/proceedings/pdf/1010.pdf).

[8] I.V. Konnov, O.A. Kashina, and E. Laitinen, "Two-level decomposition method for resource allocation in telecommunication network", *Int. J. Digit. Inf. and Wirel. Comm.*, 2012, vol.2, no. 2, pp.150–155.

[9] J.G. Kemeny and J.L. Snell, *Finite Markov Chains*, Van Nostrand, Princeton, 1960.

[10] B.T. Polyak, *Introduction to Optimization*. Nauka, Moscow, 1983; English transl. in Optimization Software, New York, 1987.

[11] I.V. Konnov, "Application of the penalty method to non-stationary approximation of an optimization problem", *Russ. Mathem. (Iz. VUZ)*, to appear.

[12] I.V. Konnov, "Application of penalty methods to non-stationary variational inequalities", *Nonl. Anal.* (submitted).

Proceedings of FORS40 - 2013 Lappeenranta, Finland - ISBN 978-952-265-435-9

Economic Growth in a Network under Mutual Dependence of Agents

Yury P. Martemyanov, Vladimir D. Matveenko

Center for Market Studies and Spatial Economics
National Research University Higher School of Economics
Moscow, Russia
ymartemyanov@hse.ru, vmatveenko@hse.ru

Abstract—We consider a dependence of the growth rate on the elasticity of factor substitution in a framework of a model of mutual dependence of n agents. This model is interpreted as a network structure and can be used to analyze agglomerations. The development is modeled as an increase in values of the agents in a dynamic system with CES functions. We investigate the cases of high and low complementarity of activities. In particular, we receive conditions allowing the identification of the cases when the elasticity of factor substitution has a positive effect on the growth rate under high complementarity of activities, and when the elasticity of factor substitution has a negative effect on the growth rate under low complementarity of activities.

Keywords— Agglomeration, Weak Links, Network Structure, Externalities, Production Functions, Growth Rate.

I. INTRODUCTION

The dependence on the growth rate on the elasticity of substitution has been studied previously in a framework of the economic growth theory (see, e.g., [1], [2]). We consider this issue in relation to a model of mutual dependence of n agents, which can be used to study dynamics of agglomerations.

Agglomeration is a complex network structure, which includes relationships among aggregated agents, such as economic sectors in a city (for instance, road networks, bridges, public transportation, housing, medicine, etc.), control structures, as well as industrial, professional and social networks which are included into the greater network as components. Methods of network analysis, which come back to [3], have become an important tool for analysis of the economy (see [4]). Research of recent decades emphasizes the importance of weak links for economic systems ([5], [6]) as well as of the higher order relationships causing a cascade effect ([7]). Economic growth rates are associated with a presence of weak links, and the latter are often associated with inadequate size of existing positive externalities. Network structures with positive externalities are particularly important in cities and agglomerations. Eigenvectors and eigenvalues has

The research is supported by the Russian Foundation for Basic Research (project 11-01-00878a).

a special place in many network models (see, e.g., [8], [4], [7]). The role of their idempotent[1] counterparts is shown by [9].

II. THE MODEL

We consider a network consisting of n agents. The state of agent i at time t is described by one number – the value of the agent, $x_i(t)$. The initial values of agents, $x_i(0)$, $i=1,...,n$, are specified. The development is modeled as an increase in the values of agents in a dynamic system with CES functions:

$$x_i(t+1) = (\sum_{j=1,...,n} \alpha_{ij}(A_{ij}x_j(t))^p)^{\frac{1}{p}}, \quad (1)$$

where $t = 0,1,...; i = 1,2,...,n; j = 1,2,...,n; \alpha_{ij} \geq 0, A_{ij} \geq 0$. As usually for production functions, it is assumed that $p \in (-\infty, 0) \cup (0,1)$. Cases $p<0$ and $p>0$ are interpreted, respectively, as cases of high and low complementarity of activities of different agents. Coefficients a_{ij} and A_{ij} can reflect, in particular a power of positive externalities created by agents j and useful for agents i. The system (1) can be written in matrix form:

$$X^p(t+1) = \tilde{A}X^p(t),$$

where $\tilde{A}$ is a square matrix with non-negative elements:

$$\widetilde{A_{ij}} = \alpha_{ij}A_{ij}^p;$$

$$X^p(t) = (x_1^p(t),...,x_n^p(t))^T$$

is the vector of degrees of states of agents in the period t; T is a sign of transposition.

A pattern of the dynamics is determined by the structure of the matrix $\tilde{A}$. We assume that this matrix is irreducible and primitive. Let $\tilde{\lambda}$ be its Frobenius eigenvalue. Then

$$\tilde{\lambda}^{-t}X^p(t) \to \tilde{X},$$

where $\tilde{X}$ is a Frobenius eigenvector. Consequently,

$$(\tilde{\lambda}^{-t})^{\frac{1}{p}}X(t) \to \tilde{X}^{\frac{1}{p}},$$

where $X(t) = (x_1(t),...,x_n(t))^T$; the number $\lambda = \tilde{\lambda}^{\frac{1}{p}}$ represents

---

[1] A binary operation $\oplus$ on a set M is called idempotent if $a \otimes a = a$ for each $a \in M$. Such are the operations max and min.

the growth factor of the model. We investigate the dependence of λ on the elasticity of substitution,

$$\sigma = \frac{1}{1-p};$$

the latter varies, obviously, in the same direction as p.

III. SOME ANALYTICAL RESULTS

3.1. If $A_{ij}=1$ for all $i,j=1,...,n$, then the matrix $\tilde{A}$ and its Frobenius eigenvalue $\tilde{\lambda}$ do not depend on p.

Proposition 1: If $A_{ij}\equiv1$ $(i=1,...,n, j=1,...,n)$ and the matrix $\tilde{A}$ provides economic growth, then: 1) the elasticity of factor substitution has a positive effect on the growth rate under high complementarity of activities; 2) the elasticity of factor substitution has a negative effect on the growth rate under low complementarity of activities.

Proof. Direction of dependence of the growth rate on the elasticity of substitution depends on the sign of the derivative

$$\lambda' = \tilde{\lambda}^{\frac{1}{p}} \ln \tilde{\lambda} \left(-\frac{1}{p^2}\right). \tag{2}$$

The presence of the growth means that $\lambda>1$. If $p<0$ (there is a high complementarity of activities), then $\tilde{\lambda} = \lambda^p < 1$, hence the right side of (2) is positive, and $\lambda' > 0$. Conversely, if $p>0$ (low complementarity), $\tilde{\lambda} > 1$, and hence $\lambda' < 0$. Q.E.D.

3.2. If $A_{ij}<1$ for all $i,j=1,...,n$, then each element of the matrix $\tilde{A}$ is decreasing in p, hence (see Debreu and Herstein, 1953) Frobenius eigenvalue $\tilde{\lambda}$ is decreasing in p, that is $\tilde{\lambda}' < 0$.

Proposition 2: If, for all $A_{ij}<1$, the matrix $\tilde{A}$ provides economic growth, then elasticity of factor substitution has a positive effect on the growth rate under high complementarity of activities.

Proof. Direction of dependence of the growth rate on the elasticity of substitution is determined by the sign of the derivative

$$\lambda' = \frac{1}{p} \tilde{\lambda}^{\frac{1}{p}-1} \tilde{\lambda}' + \tilde{\lambda}^{\frac{1}{p}} \ln \tilde{\lambda} \left(-\frac{1}{p^2}\right). \tag{3}$$

We have $\lambda>1$ (the presence of growth) and $p<0$ (high complementarity), it follows $\tilde{\lambda} = \lambda^p < 1$. Thus, each of the two terms on the right-hand side of (3) is positive, and hence $\lambda'>0$. Q.E.D.

3.3. If $\sum_{j=1}^{n} \alpha_{ij} = 1$, $i=1,...,n$, then in the limit $p\to0$ the system (1) is transformed into a system with Cobb-Douglas functions:

$$x_i(t+1) = \prod_{j=1}^{n}(A_{ij}x_j(t))^{\alpha_{ij}} = B_i \prod_{j=1}^{n}(x_j(t))^{\alpha_{ij}} \tag{4}$$

The trajectory is called balanced if

$$x_i(t+1) = \lambda x_i(t),$$

for all $i=1,...,n$, $t=0, 1, ...$ Then

$$\lambda x_i(t) = B_i \prod_{j=1}^{n}(x_j(t))^{\alpha_{ij}}.$$

Proposition 3: The trajectory of (4) converges to a balanced one.

Proof. Taking the logarithms, the system (4) can be written as follows:

$$X(t+1)=b+AX(t),$$

where $b = (\ln B_1, ..., \ln B_n)^T$; A is matrix with elements (α_{ij}), and $X(t)$ are vectors of logatithms. We express sequentially $X(\bar{t})$, $\bar{t} = 1, ..., t$:

$$X(1)=b+AX(0),$$

$$X(2)= b+AX(1)=b+Ab+A^2X(0),$$

$$X(3)=b+AX(2)=b+Ab+A^2b+A^3X(0),$$

$$...$$

$$X(t)=(E+A+A^2+...+A^{t-1})b+A^tX(0).$$

Then

$$X(t+1)—X(t)=A^t[b+(A—E)X(0)].$$

Since the matrix A is stochastic, as $t\to\infty$ the sequence of matrices A^t converges to a matrix with identical rows, hence $X(t+1)—X(t)$ converges to a vector with identical elements, we write it as $(\ln \lambda, ..., \ln \lambda)^T$. Hence,

$$\ln \frac{x_i(t+1)}{x_i(t)} \to \ln \lambda,$$

and then

$$\frac{x_i(t+1)}{x_i(t)} \to \lambda,$$

$i=1,...,n$. Q.E.D.

In the special case when $n=2$ and $\alpha_{i1} + \alpha_{i2} = 1$, we find:

$$\lambda = B_1^{\frac{1-\alpha_{22}}{2-\alpha_{11}-\alpha_{22}}} B_2^{\frac{1-\alpha_{11}}{2-\alpha_{11}-\alpha_{22}}},$$

thus, the growth rate is described by the Cobb-Douglas function of the efficiencies of the productive sectors;

$$\frac{x_2}{x_1} = \left(\frac{B_2}{B_1}\right)^{\frac{1}{2-\alpha_{11}-\alpha_{22}}},$$

thus, the higher the relative productivity of a sector is, the larger the proportion of this sector in the structure of balanced economy is.

3.4. If $\sum_{j=1}^{n} \alpha_{ij} = 1$, $i=1,...,n$, then in the limit at $p\to—\infty$ CES function (1) converges to Leontief production function, and our model is transformed into the model introduced in Matveenko, 1995:

$$x_i(t+1) = \min_{j=1,...,n} A_{ij} x_j(t).$$

In terms of tropical (idempotent) mathematic this system can be written as $X(t+1)=A \oplus X(t)$, where $A \oplus X(t)$ is a multiplication of matrices with operations $\oplus = \times$, $+ = \min$. Growth factor, λ, is an idempotent eigenvalue; the corresponding network analysis is provided in Matveenko, 1990, 2009. The growth rate can be calculated, in particular, by

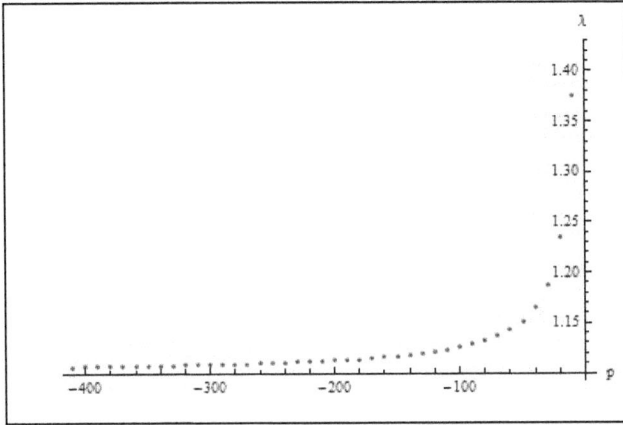

Fig. 1. The dependence of λ on p if $\alpha_{i1} + \alpha_{i2} \le 1$ and $p < 0$.

indicating an optimal subgraph; the latter is an analog of a turnpike in economic dynamic models.

IV. THE RESULTS OF NUMERICAL EXPERIMENTS

In general, analytical results are difficult to obtain, however the numerical analysis of trajectories shows patterns of dependence of the growth factor, λ, on the parameters α_{ij} and p. The nature of the dynamics of the system can radically change even with a small change in one parameter. Consider the example of matrix

$$A = \begin{pmatrix} 4 & 2 \\ 1 & 1.1 \end{pmatrix}$$

and the initial vector $X(0)$ such that if $p < 0$, then all the coordinates are non-zero, and if $p > 0$, then at least one coordinate is non-zero.

A. If $p < 0$

1) If $\alpha_{i1} + \alpha_{i2} \le 1$: λ increases in p (see Fig. 1).
2) If $\alpha_{i1} + \alpha_{i2} > 1$: λ decreases in p (see Fig. 2).

B. If $p \to -\infty$

1) If $\alpha_{i1} + \alpha_{i2} \le 1$: growth rate λ converges to the idempotent eigenvalue equal to 1.1, above (see Fig. 1).

2) If $\alpha_{i1} + \alpha_{i2} > 1$: growth rate λ converges to 1.1 below (see Fig. 2).

C. If $p > 0$

1) If $\alpha_{i1} + \alpha_{i2} \le 1$: λ increases in p (see Fig. 3).
2) If $\alpha_{i1} + \alpha_{i2} > 1$: λ decreases in p (see Fig. 4).

D. Case $p = 0$: this point is the break point of the 2nd kind.

1) If $p \to 0$— (left):
  a) If $\alpha_{i1} + \alpha_{i2} \le 1$: $\lambda \to +\infty$ (see Fig. 1).
  b) If $\alpha_{i1} + \alpha_{i2} > 1$: $\lambda \to 0$ — (see Fig. 2).
2) If $p \to 0+$ (right):
  a) If $\alpha_{i1} + \alpha_{i2} \le 1$: $\lambda \to 0+$ (see Fig. 3).
  b) If $\alpha_{i1} + \alpha_{i2} > 1$: $\lambda \to +\infty$ (see Fig. 4).

V. CONCLUSION

We investigated the dependence of the growth rate on the elasticity of factor substitution in a framework of a model of mutual dependence of n agents. We obtained some analytical and numerical results under high and low complementarity of activities. We received some conditions allowing the identification of the cases when the elasticity of factor substitution has a positive effect on the growth rate under high

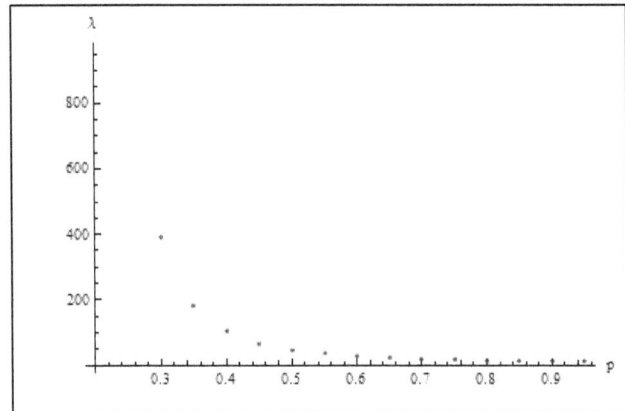

Fig. 3. The dependence of λ on p if $\alpha_{i1} + \alpha_{i2} \le 1$ and $p > 0$.

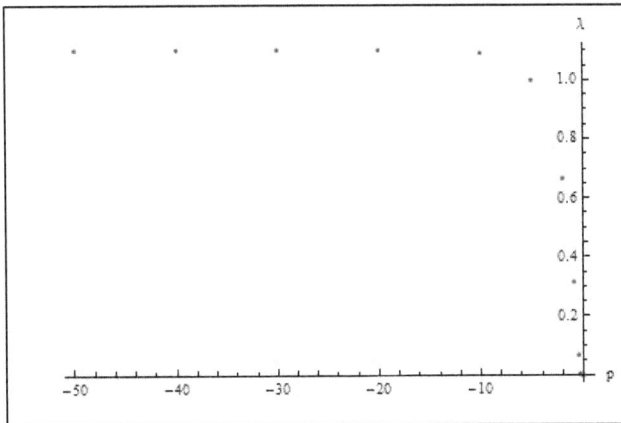

Fig. 2. The dependence of λ on p if $\alpha_{i1} + \alpha_{i2} > 1$ and $p < 0$.

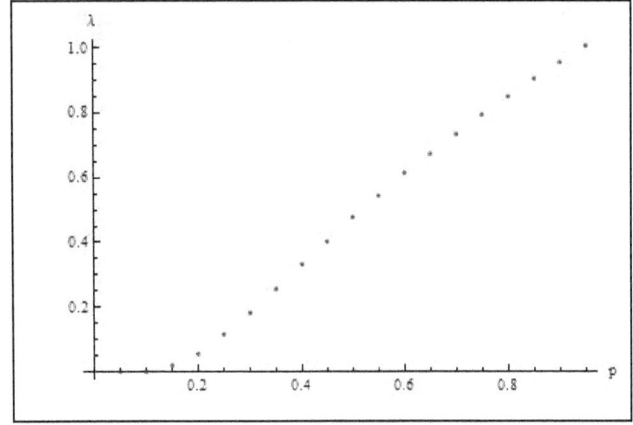

Fig. 4. The dependence of λ on p if $\alpha_{i1} + \alpha_{i2} > 1$ and $p > 0$.

complementarity of activities, and when the elasticity of factor substitution has a negative effect on the growth rate under low complementarity of activities.

REFERENCES

[1] O. de La Grandville, "In quest of the Slutsky diamond," Am. Econ. Rev, vol. 79(3), pp.468-481, 1989.

[2] T. Palivos and G. Karagiannis, "The elasticity of substitution as an engine of growth," Macroeconomic Dynamics, vol.14(5), pp.617-620, 2010.

[3] W. Leontief, The structure of American economy: 1919-1929: An empirical application of equilibrium analysis, Cambridge: Harvard University Press, 1941.

[4] M.O. Jackson, Social and Economic Networks, Princeton: Princeton University Press, 2008.

[5] M. Kremer, "The O-ring theory of economic development," Quart. Jour. Econ, vol. 108, pp.551-575, 1993.

[6] O. Blanchard and M. Kremer, "Disorganization," Quart. Jour. Econ, vol.112(1), pp.1091-1126, 1997.

[7] D. Acemoglu, V.M. Carvalho, A. Ozdaglar, and A. Tahbaz-Salehi, "The network origins of aggregate fluctuations," Econometrica, vol. 80(5), pp.1977-2016, September 2012.

[8] C. Ballester, A. Calvo-Armengol, and Y. Zenou, "Who's who in networks. Wanted: the key player," Econometrica, vol. 74(5), pp. 1403-1417, September 2006.

[9] V.D. Matveenko, "Development with positive externalities: the case of the Russian economy," Jour. of Policy Modeling, vol. 17(3), pp.207-221, 1995.

[10] G. Debreu and I.N. Herstein, "Nonnegative Square Matrices," Econometrica, vol. 21(4), pp.597-607, 1953.

[11] V.D. Matveenko, "Optimal trajectories of a dynamic programming scheme and extremal degrees of nonnegative matrices," Diskretnaya matematika, vol. 2(1), pp.59-71, 1990 (In Russian).

[12] V.D. Matveenko, "Optimal paths in oriented graphs and eigenvectors in $max-\otimes$ systems," Discrete Mathematics and Applications, vol.19(4), pp.389-409, 2009.

Multi-expert multi-criteria ranking method that uses scorecards, fuzzy heavy OWA, and similarity measures.

Pasi Luukka and Mikael Collan
School of business
Lappeenranta University of Technology
Lappeenranta, Finland
{pasi.luukka,mikael.collan}@lut.fi

Abstract—This paper proposes a new multi-expert multi-criteria fuzzy decision making method for pre-selection of acquisition targets. Experts evaluate each target by using a crisp scorecard. The crisp scorecards are aggregated to yield one fuzzy scorecard. Fuzzy heavy OWA is used to form an aggregated fuzzy scorecard score. Targets are ranked by comparing their aggregated scores to a fuzzy ideal solution by using a fuzzy similarity measure. The method is capable of preserving information that relates to the dispersion of the expert assessments of each scorecard criterion and taking it into consideration in the ranking of alternatives.

Keywords-MCDM; fuzzy heavy OWA; ranking; fuzzy similarity measure;

I. INTRODUCTION

Investments, projects, and other future-oriented assets that a firm is contemplating to start suffer often from the availability of less than perfect information. This is caused by the complexity connected to the future and the (in)ability to assess the future precisely. That is, both uncertainty and estimation imprecision play a role in decision-making. Yet under these circumstances managers are faced with the problem of selecting the best investments, projects, or other assets among many that compete for financing.

Problems of this type can be approached by involving estimates from more than one expert, to obtain more than one view of the future, by using methods that are able to consider multiple relevant criteria to give an overall understanding, and that are able to reflect uncertainty and imprecision, such as fuzzy numbers [1, 2].

This paper presents a new method for multi-agent multi-criteria ranking that uses fuzzy numbers to reflect uncertainty and imprecision and that is able to preserve the information contained within the fuzzy numbers until the final ranking of alternatives.

The method is based on information collected by using scorecards, a simple tool for collecting expert information (evaluations) on multiple criteria with regards to the alternatives under analysis. Scorecards were brought to fame

in mid-1990's by the introduction of the balanced scorecard [3-5]. The method expects that multiple experts are used to evaluate each alternative, thus resulting in multiple scorecard evaluations per alternative. These crisp evaluations are merged into one fuzzy evaluation that reflects the dispersion of the experts' evaluations and the result is in effect a fuzzy scorecard. More on fuzzy scorecards can be found in, for example [6].

The fuzzy scorecard score is aggregated by using fuzzy heavy OWA [7] to allow the ordering of criteria based on expert scores, using a totaling type operator that "finds home" in scorecards, and weighing the scores according to preset policies that allow, for example, penalization of low scores.

A simple method of comparing the fuzzy similarity of each alternative to a reference "ideal alternative" is used in the ranking of alternatives; the closer the alternative is to the "ideal alternative" the better. Using this kind of a ranking method is a new approach. Possible areas of application include, for example, human resources selection, screening of acquisition targets, and screening of funding applications.

This paper proceeds as follows: section II shortly introduces the mathematical background, section III presents the proposed method, section IV illustrates the method with a numerical example, and section V concludes.

II. MATHEMATICAL BACKGROUND

A. Preliminaries

The following is a brief review of some definitions related to the basic concepts that are essential to the proposed multi-expert multi-criteria decision making (MEMCDM) model.

Definition 1. A triangular fuzzy number A can be defined by a triplet (a_1, a_2, a_3). The membership function $\mu_A(x)$ is defined as [6]

$$\mu_A(x) = \begin{cases} 0, & x < a_1 \\ \dfrac{x - a_1}{a_2 - a_1}, & a_1 \le x \le a_2 \\ \dfrac{x - a_3}{a_2 - a_3}, & a_2 \le x \le a_3 \\ 0, & x > a_3 \end{cases} \qquad (1)$$

Definition 2. The α-cut of fuzzy number A is defined as

$$A^\alpha = \{x_i \mid \mu_A(x_i) \ge \alpha, \quad x_i \in X\} \qquad (2)$$

where $\alpha \in [0,1]$. A^α is a non-empty bounded closed interval contained in X and it can be denoted by $A^\alpha = [a_l(\alpha), a_r(\alpha)]$. For arithmetic operations for triangular fuzzy numbers we refer to [8].

B. Fuzzy Heavy OWA

An essential concept that is used in the proposed method is fuzzy OWA [9-11]; more precisely the fuzzy heavy ordered weighted averaging (FHOWA) operator introduced in [7] and defined as follows:

Definition 3. Let U be the set of fuzzy numbers. A FHOWA operator of dimension n is a mapping FHOWA: $U^n \rightarrow U$ that has an associated weighting vector W of dimension n such that the sum of the weights is between [1,n] and $w_i \in [0,1]$, then:

$$FHOWA(\hat{a}_1, \hat{a}_2, \cdots, \hat{a}_n) = \sum_{i=1}^{n} w_i b_i \qquad (3)$$

where b_i is the i-th largest of the ($\hat{a}_1, \hat{a}_2, \cdots, \hat{a}_n$), which are now fuzzy triangular numbers of form definition 1.

Fuzzy heavy OWA operator requires a way to order fuzzy numbers. For this we use the method presented by Kaufman and Gupta in [12].

A linear order can be found by using different selected properties of fuzzy numbers as criteria. Here three different criteria are used for finding an order: if the first criterion does not produce a strict linear order, then the second and ultimately the third criterion are used. The three used criteria are:

(i) Removal: Let us consider an ordinary number $k \in R$ and a fuzzy number A. Left side removal of A with respect to k, denoted by $R_l(A,k)$, is defined as the area bounded by k and the left side of the fuzzy number A. Right side removal, $R_r(A,k)$ is defined similarly. Removal of the fuzzy number A

with respect to k is defined as the mean of $R_l(A,k)$ and $R_r(A,k)$. Thus,

$$R(A,k) = \tfrac{1}{2}\big(R_l(A,k) + R_r(A,k)\big) \qquad (4)$$

The position of k can be located anywhere on the x-axis including k = 0. Areas of fuzzy numbers, by definition, are positive quantities, but here they are evaluated by integration, taking into account the position (negative, zero, or positive) of the variable x; therefore, $R(A,k)$ can be positive, negative, or null. The first criterion is removal with respect to k.

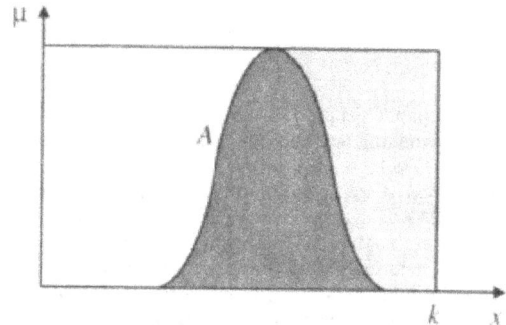

Figure 1: Removal number with regards to *k*. $R_l(A,k)$ is area of *A* + area in between *A* and *k*. $Rr(A,k)$ is area between *A* and *k*. Figure is taken from [16].

Two different fuzzy numbers can have the same removal with respect to the same k. In fact, this criterion decomposes a set of fuzzy numbers into classes that have the same removal number. If the origin 0 is conveniently moved to the left, it is possible, in this case that all of the fuzzy numbers will have positive removal numbers. Hence, the removal numbers become positive, if k is correctly chosen. The removal number with respect to a given k, therefore, can be taken as a measure of distance and can be used for the ordering of the fuzzy numbers. The removal number $R(A,k)$ defined in this criterion, relocated to k = 0 is equivalent to an "ordinary representative" of the fuzzy number. In the case of a triangular fuzzy number this ordinary representative is given by

$$\tilde{A} = \frac{a_1 + 2a_2 + a_3}{4} \qquad (5)$$

where A = (a₁, a₂, a₃).

(ii) Mode: In each class of fuzzy numbers one should look for the mode. The modes can be used to generate sub-classes. If the fuzzy numbers under consideration have a non-unique mode, one takes the mean position of the modal values. It must be noted that this is only one way of obtaining sub-classes, and one may need a third divergence criterion for further sub-classification. For the "mode" a usual choice is the core value of the fuzzy number

$$Mode(A) = \{x \in U \mid A(x) = 1\} \qquad (6)$$

Mode in the case of triangular number reduces to *Mode(A)=a₂*.

(iii) Divergence: If we consider in each sub-class, the divergence around the mode, we obtain sub-sub-classes, and this criterion may be sufficient to obtain the final strict linear ordering of fuzzy numbers.

$$Divergence(A) = sup(supp(A)) - \inf(supp(A)) \quad (7)$$

Divergence in the case of triangular number reduces to *Divergence(A)=(a₃-a₁)*.

When one orders fuzzy numbers to size order, one proceeds as follows: apply the above presented criteria in the order (i)-(ii)-(iii), such that if the strict linear order is not obtained, then move to the next criterion.

C. Scorecard, fuzzy scorecard, and FHOWA

The evaluation of the criteria is normative and based on incomplete information, due to the scarcity of information and the need to keep the screening of acquisition targets unknown to the targets themselves. This is why it makes sense to include the normative estimate of many (here three) experts in the target evaluation. The evaluation is done, by using crisp scores to indicate the best estimate of each expert. Uncertainty related to the estimates may be difficult or impossible to estimate and would potentially require a lot of time. In the proposed method a number of (in the example three) experts use a scorecard to evaluate alternatives. In Table 1 there is an example of how they apply the scorecard to evaluate alternative 1.

Table 1: Scorecard evaluation of alternative 1 from three decision makers with regards to five criteria.

Criteria	DM_1	DM_2	DM_3
C1	7	5	4
C2	7	6	5
C3	9	10	9
C4	9	9	8
C5	9	8	7

After the evaluation of each alternative a fuzzy scorecard of type [6] is created by using the three crisp evaluations to form a fuzzy triangular number of the form given in equation (1). The result is presented in Table 2.

Table 2: Fuzzy scorecard created from Table 1.

Criteria	a_1	a_2	a_3
C1	4	5	7
C2	5	6	7
C3	9	9	10
C4	8	9	9
C5	7	8	9
Sum	33	37	42

In the fuzzy scorecard the fuzzy numbers coming from the evaluation of different criteria are added together and the obtained total value is later used in decision-making. This makes the fuzzy scorecard a "totaling type of operator" that provides distinct non-redundant information about the candidate.

In fuzzy scorecard, all information available is used, as pieces of information are independent of each other. The fact that fuzzy scorecards are in some sense a "totaling type operator" makes them different from averaging type operators; if one would apply an averaging operator to fuzzy scorecards one would lose this property for providing distinct non-redundant information, and one would not be able to use (preserve) all the collected information. The preservation of this type of totaling property has been one of the main aims in developing the heavy OWA operators [13]. This makes the use of fuzzy heavy OWA [7] clearly applicable and justifiable together with fuzzy scorecards, since FHOWA is designed to be applied as a totaling type of operator for problems that cannot utilize averaging type of operators like fuzzy OWA. By applying FHOWA with fuzzy scorecards distinct, non-redundant information can be provided and all the collected information can be used.

D. Fuzzy similarity measure

For measurement of similarity between two fuzzy triangular numbers quite new method introduced in [14] is used. The method is simplified to be used with triangular fuzzy numbers in the following:

Definition 4. Let A=(a₁,a₂,a₃) and B=(b₁,b₂,b₃) be fuzzy triangular numbers and U be the set of fuzzy numbers. A similarity operator S is a mapping S: U ×U→[0,1]. Similarity function S is defined as

$$S(A,B) = \left((1-|Xa - Xb|)\right)\times \frac{\min(P(A),P(B))+ \min(A(A),A(B))}{\max(P(A),P(B))+ \max(A(A),A(B))} \quad (8)$$

where perimeters are

$$P(A) = \sqrt{(a_1 - a_2)^2 +1} + \sqrt{(a_2 - a_3)^2 +1} + (a_3 - a_1)$$
$$P(B) = \sqrt{(b_1 - b_2)^2 +1} + \sqrt{(b_2 - b_3)^2 +1} + (b_3 - b_1)$$

and areas are

$$A(A) = \frac{1}{2}(a_3 - a_1)$$

$$A(B) = \frac{1}{2}(b_3 - b_1)$$

and center-of-gravity points are

$$Xa = \frac{2Ya \times a_2 + (a_3 + a_1)(1 - Ya)}{2}, \quad Xb = \frac{2Yb \times b_2 + (b_3 + b_1)(1 - Yb)}{2}$$

$$Ya = \begin{cases} \frac{1}{3} & if \ a_1 \neq a_3 \\ \frac{1}{2} & if \ a_1 = a_3 \end{cases}, \quad Yb = \begin{cases} \frac{1}{3} & if \ b_1 \neq b_3 \\ \frac{1}{2} & if \ b_1 = b_3 \end{cases}$$

III. THE PROPOSED FUZZY MEMCDM SYSTEM FOR RANKING

The following general situation is considered, where a finite set of alternatives $A = \{A_i \mid i = 1, \cdots, m\}$ needs to be evaluated by a committee of three decision-makers $D = \{D_l \mid l = 1,2,3\}$, by considering a finite set of given criteria $C = \{C_j \mid j = 1,2,\ldots,n\}$. A decision matrix representation of performance rating of each alternative A_i is considered with respect to each criterion C_j as follows:

$$X = \begin{bmatrix} x_{11} & \cdots & x_{1n} \\ \vdots & \ddots & \vdots \\ x_{m1} & \cdots & x_{mn} \end{bmatrix} \tag{9}$$

where m rows represent m possible candidates, n columns represents n relevant criteria and x_{ij} represents the performance rating of the i-th alternative with respect to j-th criterion C_j. These ratings are obtained by using fuzzy score cards and are triangular fuzzy numbers. Triangular fuzzy numbers $R=(a,b,c)$ are formed from decision makers crisp numbers. This way fuzzy ratings x_{ij} of alternatives with respect to each criterion are now $x_{ij} = (a_{ij}, b_{ij}, c_{ij})$, where $a_{ij} = \underset{l}{min}\{x_{ijl}\}$, $b_{ij} = \underset{l}{median}\{x_{ijl}\}$, $c_{ij} = \underset{l}{max}\{x_{ijl}\}$.

The next step for the decision matrix is to form a linear scale transformation to transform the various criteria scales into comparable scales. The criteria set can be divided into a benefit criteria (larger the rating, the greater the preference) and into a cost criteria (the smaller the rating, the greater the preference). The normalized fuzzy decision matrix can be represented as

$$R = (r_{ij})_{m \times n} \tag{10}$$

where B and C are the sets of benefit criteria and cost criteria, respectively, and

$$r_{ij} = \left(\frac{a_{ij}}{c_j^{\oplus}}, \frac{b_{ij}}{c_j^{\oplus}}, \frac{c_{ij}}{c_j^{\oplus}} \right) \quad j \in B$$

$$r_{ij} = \left(\frac{a_j^{\ominus}}{c_{ij}}, \frac{a_j^{\ominus}}{b_{ij}}, \frac{a_j^{\ominus}}{a_{ij}} \right) \quad j \in C$$

where $c_j^{\oplus} = \max_i(c_{ij}), j \in B$ and $a_j^{\ominus} = \min_i(a_{ij}), j \in C$.

This normalized decision matrix is aggregated with regards to the criteria by using FHOWA:

$$R_i = FHOWA(r_{i1}, r_{i2}, \cdots, r_{in}) = \sum_{j=1}^{n} w_j b_j \tag{11}$$

where b_j is the j-th largest of the vector $(r_{i1}, r_{i2}, \cdots, r_{in})$, $w_j \in [0,1]$ and $\sum_{j=1}^{n} w_j \in [1,n]$. Required ordering is done by using the ordering method given by Kaufman and Gupta in [10] and shortly presented in the previous subsection. Also the selection of the used weighting scheme must be made.

The result is that an aggregated triangular fuzzy number is available for each alternative; the next step is to order these by using a new method of "comparison to reference ideal solution". The aggregated fuzzy numbers of each alternative and the ideal solution are mapped to a scale between [0,1], and the similarity value between the ideal solution and each alternative is calculated, by applying formula (6) as

$$S_i(A_i, A^+) = \left((1 - |Xa_i - Xa^+|) \right) \times \frac{\min(P(A_i), P(A^+)) + \min(A(A_i), A(A^+))}{\max(P(A_i), P(A^+)) + \max(A(A_i), A(A^+))} \tag{12}$$

The (preference) ordering of the alternatives is found by ranking them from highest degree of similarity to the lowest.

IV. NUMERICAL EXAMPLE

Suppose that an international company desires to acquire a firm from abroad. After a preliminary screening six target companies $A_1, A_2, \ldots, A_6$ remain for further evaluation. Five benefit criteria are considered for each target, according to the experience of the company management:

1) *Target market share (C_1)*
2) *Target operating margin (C_2)*
3) *Target technology level (C_3)*
4) *Quality of key personnel (C_4)*
5) *Share of relevant core business of whole (C_5)*

The proposed method is applied to solve the problem and computational procedure is summarized as follows:

Step 1: Multiple (in this case three) decision makers use a scorecard to evaluate the target companies with respect to all five criteria.

Step 2: A fuzzy scorecard is formed from the evaluations. The fuzzy ratings for all criteria are presented in Table 3

Table 3: Target ratings; summary for all criteria.

	C_1	C_2	C_3	C_4	C_5
A_1	(4,5,7)	(5,6,7)	(9,9,10)	(8,9,9)	(7,8,9)
A_2	(8,9,9)	(7,8,10)	(9,10,10)	(7,8,8)	(5,6,10)
A_3	(8,8,10)	(9,9,10)	(7,7,10)	(6,6,9)	(8,10,10)

A_4	(6,7,8)	(7,8,9)	(7,9,10)	(6,9,10)	(9,10,10)
A_5	(6,8,9)	(8,9,10)	(9,10,10)	(8,9,9)	(7,8,10)
A_6	(7,9,10)	(6,7,8)	(6,7,7)	(9,9,10)	(8,8,10)
A_7	(5,6,8)	(6,7,8)	(6,7,7)	(8,9,9)	(8,9,10)

Step 3: A normalized fuzzy decision matrix is constructed, see Table 4.

Table 4: Normalized fuzzy decision matrix.

	C_1	C_2	C_3	C_4	C_5
A_1	(0.4,0.5,0.7)	(0.5,0.6, 0.7)	(0.9, 0.9, 1)	(0.8, 0.9,0.9)	(0.7, 0.8,0.9)
A_2	(0.8, 0.9,0.9)	(0.7, 0.8,1)	(0.9,1,1)	(0.7, 0.8,0.8)	(0.5, 0.6,1)
A_3	(0.8,0.8,1)	(0.9,0.9,1)	(0.7,0.7,1)	(0.6, 0.6,0.9)	(0.8,1,1)
A_4	(0.6, 0.7,0.8)	(0.7, 0.8,0.9)	(0.7, 0.9,1)	(0.6, 0.9,1)	(0.9,1,1)
A_5	(0.6, 0.8,0.9)	(0.8, 0.9,1)	(0.9,1,1)	(0.8, 0.9,0.9)	(0.7, 0.8,1)
A_6	(0.7,0.9,1)	(0.6,0.7,0.8)	(0.6, 0.7,0.7)	(0.9, 0.9,1)	(0.8, 0.8,1)
A_7	(0.5,0.6,0.8)	(0.6,0.7,0.8)	(0.6, 0.7,0.7)	(0.8,0.9,0.9)	(0.8,0.9,1)

Step 3: The decision matrix is aggregated using FHOWA over all criteria, with the pre-chosen weighting vector, see Table 5. In this example the weighting vector is $w =$ [1,0.7,0.7,0.7,0.1]; when this type of a weighting vector is used, a target needs to get a very high score from one criterion, needs to perform well on three criteria, and the fifth criterion can be "improved after selection". One needs to remember that the worst targets were eliminated in the pre-screening phase. The weighting vector can be selected according to the aggregation requirements. The resulting aggregated values are visible in Table 5.

Table 5: Aggregated values for the targets by using FHOWA

Candidate	FHOWA(C_1,C_2,C_3,C_4,C_5)
A_1	(2.34,2.56,2.82)
A_2	(2.49,2.81,2.99)
A_3	(2.54,2.74,3.19)
A_4	(2.36,2.89,3.11)
A_5	(2.57,2.90,3.12)
A_6	(2.43,2.65,3.03)
A_7	(2.25,2.57,2.76)

Step 5: The aggregated fuzzy numbers are scaled to the interval [0,1] (normalized) , as in Table 6.

Table 6: Normalized fuzzy numbers for the targets.

Candidate	FHOWA(C_1,C_2,C_3,C_4,C_5)
A_1	(0.73,0.80,0.88)
A_2	(0.78,0.88,0.93)
A_3	(0.80,0.86,1)
A_4	(0.74,0.91,0.98)
A_5	(0.81,0.91,0.98)
A_6	(0.76,0.83,0.95)
A_7	(0.71,0.81,0.87)

Step 6: Creation of a (fuzzy) "ideal target" and the calculation of similarity of all fuzzy numbers in Table 6 to the ideal solution. In this case the normalized ideal solution is: $A^+ = (0.9,1,1)$ and similarity of the targets to the ideal solution is given in Table 7.

Table 7: Similarity of the targets to the ideal target

Candidate	$S(A_i,A^+)$	Order
A_1	0.81	6
A_2	0.86	2
A_3	0.85	3
A_4	0.825	5
A_5	0.89	1
A_6	0.828	4
$A7$	0.79	7

This way it can be concluded that the best acquisition target would be target number five with 0.89 similarity to the "ideal target".

V. CONCLUSIONS

This paper has presented a new multi-expert multi-criteria decision-support method that is based on crisp scorecards used by multiple (three), the use of fuzzy heavy OWA, and using a new approach based on fuzzy similarity for ranking.

The benefits of the method include the use of multiple experts to obtain multiple points-of-view for considering imprecision and uncertainty that are used for the creation of a triangular fuzzy result for each criterion. The fuzzy numbers created by integrating the opinions of multiple experts work as a consensus of the opinions of the experts. By using the resulting fuzzy scorecard we are, in fact, providing distinct, non-redundant information, and we are able to preserve all the collected information.

The fuzzy scorecard total score is aggregated with the fuzzy heavy OWA operator for the purpose of including the desirable ordering property. Furthermore the fuzzy heavy OWA is used, instead of fuzzy OWA, because fuzzy heavy OWA is a totaling type of operator that is able to preserve properties of fuzzy scorecards, which would be lost if a simple averaging operator like fuzzy OWA was used.

Last step in the approach is the ranking by comparison of the fuzzy alternatives to an ideal solution (alternative). Comparison of fuzzy numbers to an ideal solution is not a trivial task. In the comparison a new fuzzy similarity measure is chosen for the job: it able to include several different types of information carried by the fuzzy number. These include the center of gravity, the perimeter of fuzzy numbers, and the area of fuzzy numbers – all properties that carry information about the alternatives that would be lost if a simple distance measure was used in the ranking.

The proposed method is usable in many instances of decision-making that are done under uncertainty and imprecision, that require the ordering property of OWA operators, but that also require the use of a weighting vector that can reflect a selection policy, and that benefit from the use of a totaling rather than an averaging operation.

REFERENCES

[1] L. A. Zadeh, "Fuzzy Sets," *Information and Control,* vol. 8, pp. 338-353, 1965.

[2] L. A. Zadeh, "Outline of a new approach to the analysis of complex systems and decision processes," *IEEE Transactions on Systems, Man & Cybernetics,* vol. 3, pp. 28-44, 1973.

[3] R. S. Kaplan and D. P. Norton, "Putting the Balanced Scorecard to Work," *Harvard business review,* vol. 71, pp. 134-147, September-October 1993.

[4] R. S. Kaplan and D. P. Norton, "The Balanced Scorecard - Measures that Drive Performance," *Harvard business review,* vol. 70, pp. 71-80, January-February 1992.

[5] R. S. Kaplan and D. P. Norton, "Linking the Balanced Scorecard to Strategy," *California Management Review,* vol. 39, pp. 53-79, 1996.

[6] M. Collan, "Fuzzy or linguistic input scorecard for IPR evaluation," *Journal of Applied Operational Research,* vol. 5, pp. 22-29, 2013.

[7] J. M. Merigo and M. Casanovas, "Using fuzzy numbers in heavy aggregation operators," *International Journal of Information Technology,* vol. 4, pp. 177-182, 2007.

[8] M. Kaufmann and M. Gupta, *Introduction to fuzzy arithmetics: theory and applications.* New York, NY: Van Nostrand Reinhold, 1985.

[9] T. Calvo*, et al., Aggregation operators: new trends and applications.* New York: Physica-Verlag, 2002.

[10] G. Canfora and L. Troiano, "An Extensive Comparison between OWA and OFNWA Aggregation," in *8th SIGEF Conference,* Napoli, Italia, 2001.

[11] H. B. Mitchell and D. D. Estrach, "An OWA operator with fuzzy ranks," *International Journal of Intelligent Systems,* vol. 13, pp. 69-81, 1998.

[12] M. Kaufmann and M. Gupta, *Fuzzy Mathematical Models in Engineering and Management Science*: Elsevier Science Publishers B. V., 1988.

[13] R. R. Yager, "Heavy OWA operator," *Fuzzy Optimization and Decision Making,* vol. 1, pp. 379-397, 2002.

[14] J. wen, X. Fan, D. Duanmu, D. Yong, "A modified similarity measure of generalized fuzzy numbers" *Procedia Engineering,* vol. 15, pp. 2773-2777, 2011.

[15] C. T. Chen*, et al.*, "A fuzzy approach for supplier evaluation and selection in supply chan management," *International Journal of Production Economics,* vol. 102, pp. 289-301, 2006.

[16] J. Mattila, "Sumean logiikan oppikirja", Art House, 2002.

AHP based decision support tool for the evaluation of works of art - Registry of Artistic Performances

Jan Stoklasa
Dept. of Math Analysis and
Applications of Mathematics
Faculty of Science
Palacky University in Olomouc
17. listopadu 1192/12, 771 46 Olomouc
Email: jan.stoklasa@upol.cz

Jana Talašová
Dept. of Math Analysis and
Applications of Mathematics
Faculty of Science
Palacky University in Olomouc
17. listopadu 1192/12, 771 46 Olomouc
Email: jana.talasova@upol.cz

Abstract—**The paper provides a description of the mathematical model for the evaluation of creative work outcomes of Czech Art Colleges. 27 categories of works of art are defined based on 3 criteria, that are partially dependent - relevance/significance, extent and institutional and media reception. The 27 categories are then ranked using the pair wise comparison method. Subsequently Saaty's AHP is used to determine a score for each category. The proposed evaluation methodology combines evaluation based on objective criteria with peer review and suggests a possible way of solving the problem of arts evaluation for funding purposes. The concept of weak consistency is used in the model as a minimum requirement on consistency of the expertly defined matrix of preference intensities. For pair wise comparison matrices with ordered categories, the fulfillment of weak consistency can be checked during the data input phase. This way the weak consistency of pair wise comparison matrices can be achieved even for large numbers of categories - unlike the full consistency in Saaty's sense. The model is currently being used for allocation of a part of the subsidy from the state budget of the Czech Republic.**

Keywords—evaluation, arts, quality assessment, MCDM, AHP, consistency.

I. INTRODUCTION

Evaluation of works of arts is a difficult task. Individual taste and preferences are an important part of the art assessment process and as such the consensual evaluation in a group of experts is not easy to obtain. However, particularly for the purposes of funds distribution on national level, tools for the assessment of performance in the area of artistic work need to be available. Multiple criteria evaluation approach able to combine (multi) expert assessment of the quality of the output with more objective criteria proves to be justified for this purpose (see the example of [1]). The paper aims to present such a tool developed and currently used in the Czech Republic [2], [3].

In section II we introduce the necessary mathematical tools and concepts. Section III summarizes the basic ideas of the Registry of Artistic Performances (RUV in Czech) - the evaluation criteria and categories of works of arts used. Basic principles of the methodology for evaluation of works of arts in the Czech Republic is also presented here. In section IV the mathematical model used to determine the scores of each category of works of arts is described. The last section

provides discussion of the results and possible future directions for further development of the model.

II. PRELIMINARIES

Let us consider n categories we need to evaluate. The multiplicative Saaty's matrix of preference intensities S can be used to express the preferences of a group of experts among pairs of categories. The square matrix $S = \{s_{ij}\}_{i,j=1,\ldots,n}$ is required to be reciprocal, that is $s_{ij} = 1/s_{ji}$ for all $i,j = 1,2,\ldots,n$. If experts input their intensities of preferences between categories i and j, and if we assume that category i is preferred to category j or of equal importance, the elements s_{ij} are chosen from the set $\{1,2,\ldots,9\}$. Saaty (see [4], [5]) suggests linguistic descriptors to be used by the experts to express their preferences (see Table I). The elements s_{ij} of

TABLE I.   SAATY'S SCALE.

s_{ij}	linguistic meanings
1	category i is **equally important as** category j
3	category i is **slightly/moderately more important than** category j
5	category i is **strongly more important than** category j
7	category i is **very strongly more important than** category j
9	category i is **extremely/absolutely more important than** category j
2,4,6,8	correspond with the respective intermediate linguistic meanings

the matrix S are expertly defined estimations of the ratio w_i/w_j, where w_i is the evaluation of category i and w_j is the evaluation of category j. Finding the evaluations $w_1,\ldots,w_n$ of the categories based on the information provided by experts through the matrix S means finding the arguments of the minimum of expression (1).

$$\sum_{i=1}^{n}\sum_{j=1}^{n}\left(s_{ij} - \frac{w_i}{w_j}\right)^2 \qquad (1)$$

The solution to this problem for all $i = 1,\ldots,n$ can be found in the form of (2).

$$w_i = \sqrt[n]{\prod_{j=1}^{n} s_{ij}} \qquad (2)$$

Alternatively (see e.g. [6]) w_i can be obtained as the ith component of the eigenvector of S corresponding to its largest real eigenvalue $\lambda_{\max}$ (also known as the spectral radius of S).

If $s_{ij} = w_i/w_j$ for all $i, j = 1, \ldots, n$, the matrix S is fully consistent in Saaty's sense. The Saaty's consistency condition can be also formulated in the following way:

$$s_{ik} = s_{ij} \cdot s_{jk}, \text{ for all } i, j, k = 1, 2, \ldots, n. \quad (3)$$

It is well known that the condition (3) is usually not fulfilled by expertly defined matrices of larger order (particularly when linguistic descriptors from Table I are used in the input process). Saaty therefore proposes an inconsistency index CI based on the spectral radius of S by (4), where n is the order of S (in our case the number of categories).

$$CI = \frac{\lambda_{\max} - n}{n - 1} \quad (4)$$

The matrix S is considered consistent enough in Saaty's sense, if condition (5) holds, where CR is the so called inconsistency ratio and RI_n is the random inconsistency index of a matrix of intensities of preferences of order n (RI_n is obtained as an average of inconsistency indices for randomly generated reciprocal multiplicative matrices of intensities of preferences of the order n).

$$CR = \frac{CI}{RI_n} < 0.1 \quad (5)$$

Other approaches to inconsistency measurement can be found for example in [7], [8], [9] or [10], Brunelli et al. provide a comparison of various inconsistency measures in [11]. The threshold set in (5) at 0.1 is rather arbitrary, although it has been successfully used in many real world applications.

An important deficiency of many of the inconsistency measures that can be found in the literature is, apart from the arbitrary definition of acceptable consistency, the fact that the actual inconsistency can be assessed only once the matrix S is completed. However if we require the experts to input a large matrix of preference intensities, the input process might be too time consuming to be repeated in case the sufficient consistency is not achieved. Various methods of dealing with this issue can be found in the literature. Some involve adjustments of the final matrix so that it becomes consistent enough, proposals of methods of inputting incomplete matrix S and computing evaluations from such matrix with missing elements can also be found (see e.g. [12]). Hence we either repeat the input process, require additional information, change the obtained information or compute evaluations regardless of the missing information.

If we need to obtain information from a group of experts concerning their preferences on a large set of categories, obtaining additional data for changes of the resulting matrix S or inputting the whole matrix again may not be feasible in real applications. It seems reasonable to define a minimum requirement on the consistency of preferences of the experts that has to be met. The requirement should also be such that can be checked during the input phase. Such minimum consistency requirement was presented in [2], [3] as weak consistency of the matrix S. According to [2] S is weakly consistent, if for all $i, j, k \in \{1, \ldots, n\}$ the implications (6) and (7) hold.

$$s_{ij} > 1 \wedge s_{jk} > 1 \implies s_{ik} \geq \max\{s_{ij}, s_{jk}\} \quad (6)$$

$$(s_{ij} = 1 \wedge s_{jk} \geq 1) \vee (s_{ij} \geq 1 \wedge s_{jk} = 1) \implies$$
$$\implies s_{ik} = \max\{s_{ij}, s_{jk}\} \quad (7)$$

The property $s_{ik} \geq \max\{s_{ij}, s_{jk}\}$ is referred to as max-max transitivity in the literature [10]. It is easy to prove (see e.g. [2]) that a consistent matrix S (in the Saaty's sense according to (3)) is also weakly consistent. Weak consistency defined by (6) and (7) is a natural and reasonable requirement on the consistency of expert preferences. If we order the categories according to their importance before we start inputting the matrix S, the weak consistency requirement can be checked in each step of the data input phase directly by the experts, as it translates into two simple conditions - the elements of S need to be nondecreasing in each row from left to right and nonincreasing from the top downwards in each column. This allows us to obtain a weakly consistent matrix at the end of the input process and no additional modifications to the matrix are necessary for this minimum consistency condition to be met.

The quasi-ordering of categories (a transitive and complete relation) can be obtained by the pairwise comparison method. The matrix of preferences and indifferences $P = \{p_{ij}\}_{i,j=1,\ldots,n}$, where $p_{ij} = 1$ iff category i is more important than category j, $p_{ij} = 0.5$ iff categories i and j are of equal importance and $p_{ij} = 0$ iff category j is more important that category i is constructed. For all $i, j = 1, \ldots, n$ it holds that $p_{ii} = 0.5$ and $p_{ij} = 1 - p_{ji}$. The row sums $R_i = \sum_{j=1}^{n} p_{ij}$, $i = 1, \ldots, n$ can be used to determine the quasi-ordering of the categories according to their importance (the largest R_i corresponds with the most important category).

III. EVALUATION OF WORKS OF ARTS IN THE CZECH REPUBLIC

The idea of the Registry of Artistic Performances is to store information concerning the works of art created by workers (teachers) of Czech Art Colleges and Faculties, classify them into categories according to the previously stated criteria and assign scores to the categories. The sum of these scores for each Art College of Faculty is used as a measure of performance of these institutions and it is used as a basis for allocating a part of the subsidy from the state budget of the Czech Republic among them. RUV is already being used for this purpose in the Czech Republic, larger involvement of RUV in the funds distribution among Artistic Colleges and Faculties is the next natural step, as well as including also single artistic study programmes on nonartistic faculties.

For the purposes of evaluation, the art sector in the Czech Republic is divided into 7 segments - architecture, design, film, fine arts, literature, music and theatre. The works of art are evaluated according to three criteria regardless of their segment of origin. In each criterion, three different levels are distinguished (denoted by capital letters that are then used for the description of categories):

Relevance or significance of the piece

A - a new piece of art or a performance of crucial significance;

B - a new piece of art or a performance containing numerous important innovations;

C - a new piece of art or a performance pushing forward modern trends.

This criterion is assessed expertly in a peer review process that will be described further in the text and plays the role of a quality assessment criterion in the model. Each segment of art provided real-life (historical) examples for levels A, B and C and also the general linguistic specifications are customized for each segment and made available to the expert evaluators.

Extent of the piece

K - a piece of art or a performance of large extent;

L - a piece of art or a performance of medium extent;

M - a piece of art or a performance of limited extent.

The levels of this criterion are again specified linguistically, by examples or by measurable for each segment on such a level of accuracy that most of the ambiguity in categorizing works of art according to this criterion is removed. This can be seen as reflecting the amount of work needed to produce the piece, the costs associated with it, number of people involved in the creation of the piece and so on. Respective units can be used to determine the levels of this criterion quantitatively.

Institutional and media reception/impact of the piece

X - international reception/impact;

Y - national reception/impact;

Z - regional reception/impact.

For this criterion lists of institutions corresponding to level X, Y and Z are provided by each segment.

Based on the levels of the three criteria 27 categories of works of art are defined. Each category is characterized by a triplet of letters (one for each criterion reflecting the level of the respective criterion) e.g. AKX, BLX, or CMZ. Each of these 27 categories is assigned a score (the mathematical model used to determine cores is described in section IV). For the purposes of the development of the mathematical model, each segment provided a list of typical works of art for each of the 27 categories. These real life examples can also be used in the peer review process by the reviewers.

In the first step the evaluation procedure involves the assessment of each work of art by its creator and his/her university or faculty. By self assessment the author proposes the initial classification of the respective work of art - assigning an initial triplet of letters. The proposed category has to be approved by the faculty of College the creator is employed at. In the second step, this evaluation is assessed by the council of the respective segment of art and either confirmation of this evaluation is issued or a second evaluation (second triplet representing the opinion of the council) is assigned. This information (either one confirmed classification of the piece of art, or two conflicting classifications) is then provided to independent reviewers in a way so that each piece of art is assessed by at least two such reviewers. The final evaluation is determined as a majority opinion of all the parties (that is as a majority based consensus on the levels of the three criteria). In indecisive cases the external (independent) evaluation is favoured.

IV. PROPOSED MATHEMATICAL MODEL

It is quite obvious that the three criteria used for the evaluation are not independent. In this case the classical approach proposed by Saaty in his AHP - that is to determine

Category	Relevance or significance	Extent	Institutional reception	Eigenvector method	Geom. means method
AKX	crucial significance	large	international	305	305
AKY	crucial significance	large	national	259	254
AKZ	crucial significance	large	regional	210	217
ALX	crucial significance	medium	international	191	194
AMX	crucial significance	limited	international	174	171
ALY	crucial significance	medium	national	138	138
ALZ	crucial significance	medium	regional	127	124
BKX	containing numerous important innovations	large	international	117	112
AMY	crucial significance	limited	national	97	94
AMZ	crucial significance	limited	regional	90	87
BKY	containing numerous important innovations	large	national	79	75
BKZ	containing numerous important innovations	large	regional	66	66
BLX	containing numerous important innovations	medium	international	62	61
BMX	containing numerous important innovations	limited	international	48	50
BLY	containing numerous important innovations	medium	national	44	46
BLZ	containing numerous important innovations	medium	regional	40	41
BMY	containing numerous important innovations	limited	national	37	38
BMZ	containing numerous important innovations	limited	regional	31	30
CKX	pushing forward modern trends	large	international	26	26
CLX	pushing forward modern trends	medium	international	24	24
CKY	pushing forward modern trends	large	national	19	20
CKZ	pushing forward modern trends	large	regional	17	18
CMX	pushing forward modern trends	limited	international	16	16
CLY	pushing forward modern trends	medium	national	12	13
CLZ	pushing forward modern trends	medium	regional	10	11
CMY	pushing forward modern trends	limited	national	9	9
CMZ	pushing forward modern trends	limited	regional	8	9

Fig. 1. Ordering of the categories, scores assigned by the Eigenvector method and by the Geometrical means method (2).

the weights of each criterion and the weights of each level within each criterion and based on these to calculate the evaluation of the categories - is not appropriate. In this case it is better to expertly compare the 27 categories using the pair wise comparison matrix. Saaty's recommendation is to split large problems (we have now a 27x27 matrix) into several problems of lower dimension. To do so would, however, in this case require the experts to express their preferences concerning really vaguely defined supercategories of works of art. This is obviously a problem, as the experts are unable to provide relevant information of this kind.

We have therefore based the mathematical model for determining scores of the 27 categories on a matrix of intensities of preferences of the order 27. The resulting matrix was required to be weakly consistent. To achieve this, the first step was to determine the quasi-ordering of the 27 categories using the pair wise comparison method. The resulting ordering in presented in the first column in Fig. 1.

In the second step the experts from various segments of art were asked to input the matrix of intensities of preferences. Having the categories ordered according to their importance resulting form the first step, the experts were instructed to provide their preferences and not to violate the weak consistency. This way a weakly consistent matrix of preference intensities was obtained (see Fig. 2), in this case $CR = 0.1996$. The scores of the categories were computed using the Eigenvector method and the Geometrical means method (2). The results are presented in Fig. 1.

V. CONCLUSION

In this paper we have presented the mathematical model underlying the Registry of Artistic Performances used to determine the scores for 27 categories of works of art. The model is

	AKX	AKY	AKZ	ALX	AMX	ALY	ALZ	BKX	AMY	AMZ	BKY	BKZ	BLX	BMX	BLY	BLZ	BMY	BMZ	CKX	CLX	CKY	CKZ	CMX	CLY	CLZ	CMY	CMZ
AKX	1	5	5	5	5	5	5	5	5	5	5	5	5	5	5	5	7	7	9	9	9	9	9	9	9	9	9
AKY		1	5	5	5	5	5	5	5	5	5	5	5	5	7	7	7	9	9	9	9	9	9	9	9	9	9
AKZ			1	3	5	5	5	5	5	5	5	5	5	5	7	7	7	9	9	9	9	9	9	9	9	9	9
ALX				1	3	5	5	5	5	5	5	5	5	5	5	7	7	9	9	9	9	9	9	9	9	9	9
AMX					1	5	5	5	5	5	5	5	5	5	5	5	5	7	7	7	9	9	9	9	9	9	9
ALY						1	3	5	5	5	5	5	5	5	5	5	5	5	5	5	7	7	7	9	9	9	9
ALZ							1	3	5	5	5	5	5	5	5	5	5	5	5	5	5	5	5	5	7	7	7
BKX								1	5	5	5	5	5	5	5	5	5	5	5	5	5	5	5	5	7	7	7
AMY									1	3	5	5	5	5	5	5	5	5	5	5	5	5	5	5	7	7	7
AMZ										1	5	5	5	5	5	5	5	5	5	5	5	5	5	5	7	7	7
BKY											1	5	5	5	5	5	5	5	5	5	5	5	5	5	7	7	7
BKZ												1	3	5	5	5	5	5	5	5	5	5	5	5	7	7	7
BLX													1	5	5	5	5	5	5	5	5	5	5	5	7	7	7
BMX														1	3	5	5	5	5	5	5	5	5	5	7	7	7
BLY															1	3	5	5	5	5	5	5	5	5	7	7	7
BLZ																1	3	5	5	5	5	5	5	5	7	7	7
BMY																	1	5	5	5	5	5	5	5	7	7	7
BMZ																		1	5	5	5	5	5	5	5	5	5
CKX																			1	3	5	5	5	5	5	5	5
CLX																				1	5	5	5	5	5	5	5
CKY																					1	3	5	5	5	5	5
CKZ																						1	3	5	5	5	5
CMX																							1	5	5	5	5
CLY																								1	3	3	3
CLZ																									1	3	3
CMY																										1	3
CMZ																											1

Fig. 2. Saaty's matrix of preference intensities as provided by the experts from various segments of arts. Consensus through all segments was required.

based on Saaty's AHP. The notion of weak consistency was introduced as a minimum requirement on the consistency on such a large matrix. By ordering the categories according to their importance prior to asking the experts to input their intensities of preferences, the weak consistency condition fulfillment can be checked in each step. After obtaining the weakly consistent matrix of preference intensities, scores of categories were computed using the Eigenvector and the Geometrical means method. The results of these two methods are similar.

The evaluation methodology presented in this paper not only deals with the difficult task of evaluation of works of art, it also provides a means for integrating the peer review component into the process of evaluation and combining it with classical evaluation criteria. RUV is currently being used in the Czech Republic for the distribution of a part of the subsidy from the state budget among Art Colleges and Faculties, next year it should open to all universities that, even partially, provide education in the field of arts.

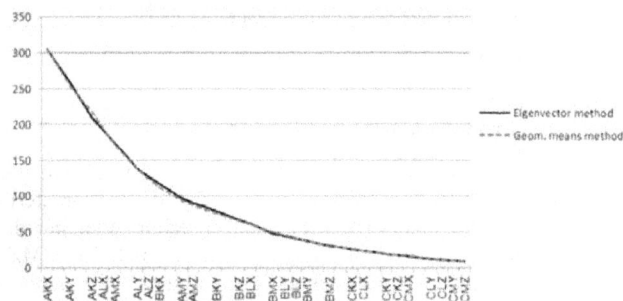

Fig. 3. Comparison of the resulting scores of categories using the Eigenvector method and the Geometrical means method.

ACKNOWLEDGMENT

This research was supported by the Centralized Developmental Projects C41 (2011) and C1 (2012) entitled *Evaluating Creative Work Outcomes to determine VKM and B3 Indexes for Art Colleges and Faculties - A Pilot Project* and financed by the Czech Ministry of Education.

REFERENCES

[1] *Smernica č. 13/2008-R zo 16. októbra 2008 o bibliografickej registrácii a kategorizácii publikačnej činnosti, umeleckej činnosti a ohlasov.* Ministerstvo školstva Slovenskej Republiky.

[2] J. Stoklasa, V. Jandová, J. Talašová, Weak consistency in Saaty's AHP - evaluating creative work outcomes of Czech Art Colleges, *Neural network world*, 23, 2013, pp. 61 - 77.

[3] J. Talašová, J. Stoklasa, A model for evaluating creative work outcomes at Czech Art Colleges, *Proceedings of the 29th International Conference on Mathematical Methods in Economics 2011 - part II, 2011, Praha, Czech Republic*, Praha, 2011, pp. 698-703.

[4] Saaty, T. L., Vargas, L. G.: *Decision Making with the Analytic Network Process: Economic, Political, Social and Technological Applications with Benefits, Opportunities, Costs and Risks.* Springer, New York, 2006.

[5] Saaty, T.L.: *The Fundamentals of Decision Making and Priority Theory with the Analytic Hierarchy Process.* Vol. VI of the AHP Series, 478 pp., RWS Publ., 2000.

[6] Saaty, T.L.: Relative Measurement and Its Generalization in Decision Making, Why Pairwise Comparisons are Central in Mathematics for the Measurement of Intangible Factors - The Analytic Hierarchy/Network Process, *RACSAM*, Vol. 102, No. 2, 2008, pp. 251 - 318.

[7] Alonso, J.A., Lamata, M.T.: Consistency in the Analytic Hierarchy Process: A New Approach, *International Journal of Uncertainty, Fuzziness and Knowledge-Based Systems*, Vol 14., No. 4, 2006, pp. 445-459.

[8] Lamata, M.T., Pelaez, J.I.: A Method for Improving the Consistency Judgemets, *International Journal of Uncertainty, Fuzziness and Knowledge-Based Systems*, Vol 10, No. 6, 2002, pp. 677686.

[9] Ji, P., Jiang, R.: Scale Transitivity in the AHP, *The Journal of the Operational Research Society*, Vol. 54, No. 8, 2003, pp. 896-905.

[10] Herrera-Viedma, E., Herrera, F., Chiclana, F., Luque, M.: Some issues on consistency of fuzzy preference relations, *European Journal of Operational Research*, 154, 2004, 98-109.

[11] Brunelli, M., Canal, L., Fedrizzi, M.: Inconsistency indices for pairwise comparison matrices: a numerical study. *Annals of Operations Research*, 2011.

[12] Fedrizzi, M., Giove, S.: Optimal sequencing in incomplete pairwise comparisons for large-dimensional problems, *International Journal of General Systems*, Vol. 42, No. 4, 2013, 366-375.

Multiple objectives and system constraints in double-deck elevator dispatching

Janne Sorsa, Mirko Ruokokoski
KONE Corporation
FI-02150 Espoo, Finland
Email: Janne.Sorsa@kone.com, Mirko.Ruokokoski@kone.com

Abstract—The elevator group control dispatches elevators to serve passenger calls. The dispatching decisions are determined by solving the Elevator Dispatching Problem (EDP). Passengers enter their destination calls with the destination control already in the lobby. In turn, the group control assigns the serving elevator immediately and announces it to the passenger. A double-deck elevator consists of two elevator cars that are attached on top of each other thus serving two adjacent floors simultaneously. In the existing double-deck destination controls, both the serving elevator and deck are fixed at once after the passenger call and not changed afterwards. In this paper, the effect of these restrictions on the solutions of the double-deck elevator dispatching problem are studied by depicting its objective space with two conflicting objectives, namely, passenger waiting and transit time. This paper aims at characterizing the relationship between passenger waiting time in the lobby and transit time inside the car as optimization objectives.

I. INTRODUCTION

The main task of an elevator group control is to dispatch the elevators under its control to passenger calls. The group control optimizes dispatching decisions by solving the Elevator Dispatching Problem (EDP). Nowadays all major elevator companies offer an advanced elevator product, *Destination Control* (DC), in which passengers enter their destination calls already in the lobby using a numeric keypad. After registering the call, the group control immediately assigns the elevator to serve the call and announces it on the screen associated with the keypad. All the existing DC products are designed according to this principle.

A double-deck elevator is a special elevator consisting of two elevator cars. The cars are attached one on top of the other and move as a single unit, thus serving adjacent floors at the same time. To allow simultaneous service of the decks on the ground floor, the lobby is arranged on two floors which are interconnected by escalators [1]. In this arrangement, the lower deck serves only the lower lobby level, while the upper deck serves only the upper lobby level. However, both decks can serve all the upper floors, the topmost floor being a possible exception to the rule if the building does not have adequate space above it for the upper deck.

Modern double-deck group controls can minimize passenger waiting and journey times as optimization objectives of the Double-Deck Elevator Dispatching Problem (DD-EDP) [2][3]. However, these objectives result in different solutions even in simple situations, which raises the foreseeability of conflicting objectives. Therefore, the DD-EDP is put here in the context of multi-objective optimization, where passenger waiting and

transit times are considered simultaneously. Transit time is considered as the second objective instead of journey time because journey time is about the same as waiting plus transit time [4]. Multi-objective optimization has not been considered earlier for different passenger service objectives, but balancing of passenger service and energy consumption has been realized [5][6].

The existing double-deck DC assigns both the serving elevator and deck immediately after the passenger call is registered [3]. Recently, a new approach to DC has been proposed, which postpones the final elevator assignment to a later time. The final moment for fixing the assignment can be, for example, when the elevator starts to decelerate to the call floor. Such a product would require a different user interface and signalization concept compared to existing products. The delayed assignment does not affect the structure of the DD-EDP but only the complexity of the actual problem instances. In this paper, objective spaces of small and moderately large DD-EDP instances are studied by relaxing the system constraints gradually. The cases under consideration are:

1) the existing DC, which considers all the existing calls as constraints and makes the immediate assignment of both the elevator and the deck only once for a newly registered call;
2) *Semi-Continuous* (SC) DC, which makes the final elevator shaft assignment immediately but allows the change of the serving deck until a defined distance from the call floor;
3) *Fully-Continuous* (FC) DC, which allows the change of both the serving elevator and the deck until a defined distance from the call floor.

In the above terms, "continuous" refers to the gradual change of the DC, which traditionally operates according to the immediate assignment policy, towards the continuous assignment policy, which is typical of the conventional control with up and down call buttons as call-giving devices. Since DC is generally less efficient in mixed lunch-hour traffic than in morning up-peak [7], the delayed elevator assignment is expected to reduce passenger service times during that time. Therefore, the example instances in this paper are taken from lunch-hour traffic simulations conducted with the Building Traffic Simulator (KONE BTS™) [8]. The instances are solved in an off-line environment with the genetic algorithm [9] adapted to double-deck elevators [2][3].

II. DD-EDP Objective Functions

The DD-EDP can be decomposed into a generalized assignment problem and a set of elevator routing problems following the well-known decomposition of the vehicle routing problem. In the case of the DD-EDP, a deck of an elevator corresponds to a vehicle and the main decision variables are deck to passenger call assignments. The set S consists of all the origin floors of the destination calls requiring pick-up, and the set T contains all the destination floors of the passengers travelling inside the elevators as well as the passengers still waiting for pick-up. The waiting time of a passenger $i \in S$ is simply the sum of the estimated arrival time of an elevator, t_i, and the time elapsed since the registration of the call, γ_i. The transit time of a passenger $i \in T$ is then the difference between the elevator arrival times to the destination and the origin floor.

Each call is also associated with the number of passengers requiring service, ω_i, which is assumed to equal one in this paper. However, in the real DC, passengers can give a group call in which they define the number of travelling passengers. The number of passengers can also be estimated from traffic statistics collected by the group control [10].

The objective functions of the DD-EDP are f_1 to minimize passenger waiting time and f_2 to minimize passenger transit time,

$$f_1 (S,T) = \sum_{i \in S} (\gamma_i + t_i) \omega_i, \tag{1}$$

$$f_2 (S,T) = \sum_{i \in T} (t_i - t_j) \omega_i, \tag{2}$$

where $j \in S$ is the known origin floor of the destination call and $i \in T$ is the destination floor of the destination call. For passengers already travelling inside the car, t_j is the actual time of pick-up as recorded by the group control.

III. Analysis of Two DD-EDP Instances

Two instances of the DD-EDP are analyzed to describe its characteristics as a multi-objective optimization problem. To get a realistic initial state and calls for the instances, an elevator group consisting of five double-deck elevators was simulated with the BTS for a building with 19 floors including a two-floor lobby on the ground floor. The initial states of the instances occurred during lunch-hour traffic simulations, in which the passenger arrival rates were 6% and 12% of the building population in five minutes, or 95 and 190 persons in absolute terms. The simulated traffic was split to incoming, outgoing, and inter-floor traffic components with proportions 40%, 40% and 20%, respectively. During the simulations, the locations, loads and other state variables of the elevators as well as the passenger calls were recorded into log-files for all solved DD-EDP instances. A single instance from both simulations was chosen to represent a small instance and a moderately large instance. The small instance included 10 passenger calls and the larger one 30 passenger calls.

A. Off-line Environment to Solve the Instances

A custom-built software was used to read the initial state and calls of an instance from the log-files. It then modified the DD-EDP system constraints and solved the problem off-line by the same genetic algorithm as used in the actual

TABLE I.   Total number of alternative and evaluated chromosomes

	Instance with 10 passengers		Instance with 30 passengers	
	Alternatives	Evaluated	Alternatives	Evaluated
DC	10	10	10	10
SC	320	297	$1.31 \cdot 10^6$	7802
FC	$6.25 \cdot 10^8$	33072	$2.44 \cdot 10^{26}$	42737

elevator product [9][2][3]. However, in this off-line study, the number of chromosomes were increased to 400 and the same value was used for the maximum number of generations. With these parameters, the genetic algorithm produces many more iterations and unique chromosomes than the actual real-time algorithm which gives a better idea of the whole objective space.

The used genetic algorithm optimizes only a single objective. However, it was originally developed to be capable of evaluating two or more objective functions and then calculating the final fitness as a weighted sum over the objective function values. Thus, the DD-EDP can be formulated as a scalarized multi-objective optimization problem with weight factor w,

$$\min w f_1 (S,T) + (1-w) f_2 (S,T). \tag{3}$$

For this study, the individual objective function values were stored to analyze the objective space afterwards. It turned out that the weighing had a strong influence on the areas where the genetic algorithm solutions converged. Therefore, instances with a large number of possible solutions, namely, the ones with both SC and FC, were solved several times with different weight factors, $w \in \{0, 0.25, 0.5, 0.75, 1\}$.

B. DD-EDP Complexity

The complexity of the DD-EDP varies a lot depending on the system constraints. Theoretically, the total number of solution alternatives equals $(2E)^C$ where E denotes the number of elevators, C the number of calls subject to optimization, and 2 stands for the two decks of the elevator. However, in the case of DC, usually $C = 1$, but in the case of FC, C equals the total number of non-served passenger calls. The case of SC differs slightly from the others since it contains usually only one new call which can be served by all elevators and $C - 1$ earlier calls for which the serving deck can be changed. Therefore, the number of solutions for SC equals $2E \cdot 2^{C-1}$. Table I shows how quickly the number of alternative solutions increases depending on the number of passenger calls and the system constraints.

The maximum number of calls at any given time is restricted by elevator capacity since the DC is capable of planning elevator routes for 1.5 roundtrips ahead of their current location. Let θ denote the maximum number of calls that can be assigned to one deck at a time. Then, group control can assign up to 2θ calls for a single elevator during one trip upwards or downwards. By combining these for the whole group, the maximum number of calls is estimated at $C_{\max} = 6E\theta$. For the elevator group used in these examples $C_{\max}$ equals 420 if 14 passenger calls is the momentary maximum along the elevator routes.

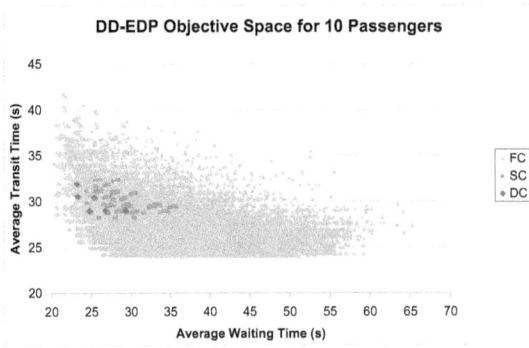

Fig. 1.   DD-EDP solutions in the objective space

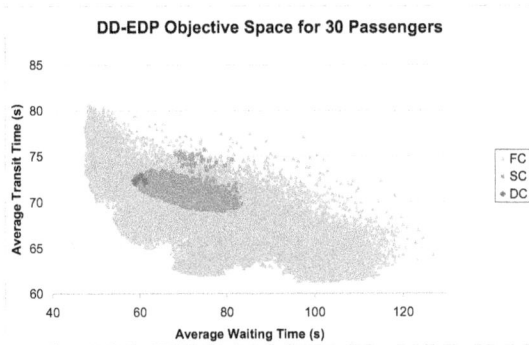

Fig. 2.   DD-EDP solutions in the objective space

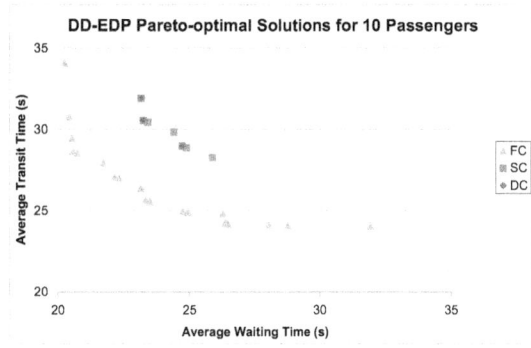

Fig. 3.   DD-EDP Pareto-optimal solutions for 10 passengers

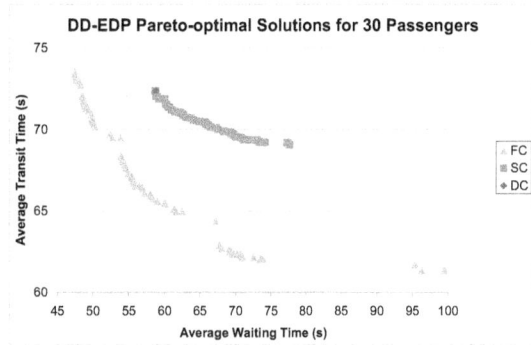

Fig. 4.   DD-EDP Pareto-optimal solutions for 30 passengers

C. DD-EDP Objective Space

Figures 1 and 2 show the evaluated solutions of the small and the large instance, and for DC, SC and FC, respectively. In the figures, each data-point corresponds to one solution to the problem instance and it depicts average passenger transit time as a function of average passenger waiting time. The solutions for SC and FC clearly exhibit a typical form for conflicting objectives since the shape of the Pareto-optimal front is visible. Since DC has only a few alternative solutions, the relationship is not well visible but seems similar to the others.

Figures 3 and 4 show the respective Pareto-optimal fronts for the small and the large problem. The Pareto front of the large instance seems to be non-continuous at some specific areas of the objective space for FC. The gaps result from the chosen weight factors that make the search to converge on the areas close to the gaps but not reaching them. The DC has only three Pareto-optimal solutions to the small instance and two for the large. This occurs probably because of the many constraints posed by the existing passengers on the service of the new one. The Pareto-optimal solutions of SC are slightly better than the solutions of the DC as there are additional possibilities to change the serving deck. According to the figures, FC promises a clearly better service level compared to DC and SC: the freedom of choice improves the minimum waiting and transit times up to 15%. However, the improvement only reflects the estimation inside the group control and its optimization model. The result can only be interpreted as an indication, and not as an actual improvement in passenger service level.

TABLE II.   PROPORTION OF EVALUATED SOLUTIONS INSIDE THE BOX BOUNDED BY IDEAL AND NADIR VECTORS, 10 PASSENGERS

	Ideal		Nadir		Inside the box
	f_1	f_2	f_1	f_2	
DC	23.16	28.98	24.72	31.91	30.0%
SC	23.16	28.26	25.88	31.91	12.8%
FC	20.25	24.04	31.93	34.09	19.0%

TABLE III.   PROPORTION OF EVALUATED SOLUTIONS INSIDE THE BOX BOUNDED BY IDEAL AND NADIR VECTORS, 30 PASSENGERS

	Ideal		Nadir		Inside the box
	f_1	f_2	f_1	f_2	
DC	58.62	72.06	61.28	72.37	50.0%
SC	58.62	69.05	77.68	72.37	94.4%
FC	47.37	61.34	99.47	78.29	79.9%

D. DD-EDP Pareto-Optimal Region

The Pareto-optimal region is a box in the objective space which is bounded by the ideal and the nadir objective vectors [11]. The ideal objective vector defines the lower left corner of the box as the minimum values of both of the objectives. The nadir point lies in the opposite corner and can in practice be defined by the values of the other than the minimum component of the ideal vector. These special objective vectors are shown in Tables II and III for the small and the large instance, respectively. As can be expected, the volumes of the Pareto-optimal regions are small for the DC, a bit wider for the SC, and the largest for the FC.

The tables also show the percentage of the evaluated solutions that lie in the Pareto-optimal region. In the case of DC, the percentage is not statistically significant because there are only 10 solutions to the problem instances. In the case of the small instance, the percentages are rather low due to tight boundaries of the Pareto-optimal region. With the large problem instance, the percentages are high but also the Pareto-optimal regions are quite large, especially for the FC. The genetic algorithm does unnecessary work when evaluating the solutions outside the Pareto-optimal region. In addition, since the volume of the Pareto-optimal region as well as the amount of useless processing varies, the algorithm should be capable of dynamically recognizing the iterations in the areas far from the Pareto front and adapting its search direction accordingly.

IV. Conclusion

In this paper, the Double-Deck Elevator Dispatching Problem (DD-EDP) was described as a multi-objective optimization problem. Generally, passenger waiting and transit times in the elevator system are conflicting objectives. Between these two objectives, a rather typical Pareto-optimal front was found by studying the solutions of two DD-EDP instances in the objective space. It seems that the more solution alternatives there are, the wider they spread in the objective space. In addition, sometimes a large proportion of evaluated solutions is inferior to the Pareto-optimal region. An on-line multi-objective optimization algorithm could use additional heuristics to avoid such bad solutions since time is always a critical resource in a real-time control system.

The system constraints of the DD-EDP arise from the Destination Control (DC), which traditionally makes the final decision of both the serving elevator and deck immediately after the passenger call is registered. The semi-continuous double-deck DC allows the group control to change the serving deck after the initial assignment but the fully continuous DC also allows the serving elevator to be changed. The fully continuous DC also requires a new user interface concept. Based on the Pareto-optimal solutions to the studied problem instances, it seems that the FC could improve passenger waiting and/or transit times by about 10-15%. However, this estimation needs to be verified by proper simulations to get a more reliable estimate.

Naturally, the question arises about what the group control should optimize. The Pareto-optimal solutions showed a rather wide range of values in both the average waiting and transit times. The addition of energy consumption as the third objective would complicate the matter even further. Traditionally, the preference order of the objectives has been the decision maker's dilemma in multi-criteria decision making. In an elevator group control, the decision making must be automatic and adaptive since it is executed continuously. However, building managers and even individual users could have a possibility to state their preferences, which the group control could take into account when making the dispatching decisions.

Acknowledgment

The authors would like to thank Professor Kalyanmoy Deb and Rick Barker, Principal of Barker Mohandas, USA, for inspiring discussions about our common interests.

References

[1] J. Fortune, "Modern double-deck elevator applications and theory," *Elevator World*, vol. 44, no. 8, pp. 63–68, 1996.

[2] J. Sorsa, M.-L. Siikonen, and H. Ehtamo, "Optimal control of double-deck elevator group using genetic algorithm," *International Transactions in Operational Research*, vol. 10, no. 2, pp. 103–114, 2003.

[3] J. Sorsa and M.-L. Siikonen, "Double-deck destination control system," *Elevatori*, vol. 37, no. 5, pp. 42–57, 2008.

[4] G. Barney, "Towards agreed traffic design definitions," *Elevator World*, vol. 53, no. 2, p. 108, 2005.

[5] T. Tyni and J. Ylinen, "Evolutionary bi-objective optimisation in the elevator car routing problem," *European Journal of Operational Research*, vol. 169, no. 3, pp. 960–977, 2006.

[6] S. Kobori, M. Iwata, N. Suzuki, and S. Yamashita, "Energy-saving techniques of elevator group control system," in *Elevator Technology 18*, A. Lustig, Ed. IAEE, 2010, pp. 167–176.

[7] J. Sorsa, H. Hakonen, and M.-L. Siikonen, "Elevator selection with destination control system," *Elevator World*, vol. 54, no. 1, pp. 148–155, 2006.

[8] M.-L. Siikonen, T. Susi, and H. Hakonen, "Passenger traffic flow simulation in tall buildings," *Elevator World*, vol. 49, no. 8, pp. 117–123, 2001.

[9] T. Tyni and J. Ylinen, "Genetic algorithms in elevator car routing problem," in *Proceedings of the Genetic and Evolutionary Computation Conference (GECCO-2001)*. San Francisco, CA, USA: Morgan Kaufman Publishers, 2001, pp. 1413–1422.

[10] M.-L. Siikonen, "Planning and control models for elevators in high-rise buildings," Ph.D. dissertation, Helsinki University of Technology, Systems Analysis Laboratory, 1997.

[11] K. Deb, *Multi-Objective Optimization using Evolutionary Algorithms*. Chichester, England: John Wiley & Sons, 2001.

Scheduling and Routing in Home Health Care Service

Thi Viet Ly Nguyen and Roberto Montemanni
Dalle Molle Institute for Artificial Intelligence (IDSIA - USI/SUPSI)
Galleria 2, 6928 Manno, Switzerland
Email: {vietly, roberto}@idsia.ch

Abstract—**In this paper we address the home health care services problem in terms of routing and scheduling. The aim of the study is to determine a feasible working plan for nurses in order to offer patients the best possible solution in terms of quality of service and economy while satisfying the demands of patients and nurses as well as the related constraints. Besides giving a brief overview of related literature, we describe a new extended version of the existing home health case service problems and propose a mixed integer linear programming formulation. Computational results conducted based on a set of randomly generated home health case scenarios reveal that the proposed model is flexible and applicable in practice.**

I. INTRODUCTION

One of the vital topics in modern societies is the improvement of life expectancy. The improvement of the quality of medical and paramedical care services taken place at the house of patients can play an important factor in this context. Consequently the problem has caught the attention of researchers and organizations during the last decade and has been considered as a home health care (HHC) service presenting a substitute to the conventional hospitalization to meet the increasing demands of patients who are discharged from hospital and the elderly who are in need of support and medical treatment for keeping the best possible clinical conditions, which is a vital aspect to help them recover completely and quickly. In order to provide the best HHC service for residents, the care providers are facing a number of challenging issues, such as service labor-intensiveness, growing operation costs, increasing demand for satisfaction level of both patients and workforce associations, and the regulations regarding working time. Therefore, in attempt to satisfy all the desires of patients, nurses, and care providers within a limit operation cost it is worth considering all the operational decisions affecting the efficiency of the HHC service within a same optimization framework. Among these decisions, assigning nurses to patients, scheduling the treatments, and routing the tours of nurses are the most important and challenging decisions.

These decisions have been addressed in some recent publications under a variety of assumptions, constraints and objective functions. For instance, some studies deal with the scheduling horizon of one day [1], [2], while others consider a longer scheduling horizon of up to one week [3], [4], [5]. The cost of overtime is taken into account in [4], [1] but is ignored in [3], [5], [2]. Almost the publications assume that all nurses have to begin and end their working day at a central office, however, the multiple central offices are stated in [5]. Some problem-specific assumptions are also considered to make the model of HHC service more realistic. In particular, qualification requirements are often taken into account ([3], [1], [5], [2]): each treatment (injections, taking blood pressure, taking temperature, blood transfusion, different types of medication, etc) requires an exact level of skill and each nurse provides a certain skill level. These studies assume a nurse can be able to perform a required treatment only if that nurse has a skill which is at least equal to the required skill, implicitly implying a clear ranking of the skills that is certainly not obvious in practice. In order to overcome this limitation, [4] assume that each nurse provides a set of skill and a nurse can only work the requests which require the skills are the same as her skills. In [3], [1] we can see unsatisfied requests are allowed with and without considering a penalty cost. However, in [4] all the unsatisfied requests are filled up with a *dummy* nurse who is not restricted to any constraints. Then, a penalty cost is added into the objective function.

A set of operation days of associated request is the important information to create the working plan for the nurses. While some studies consider it as given information, [3] propose three policies to generate a set of possible combinations of days, e.g., Monday-Friday or Wednesday-Tuesday-Thursday. The possible combinations are called patterns and are used to operate the patient requests. The impact on the quality of the solutions of the policies used to generate these patterns is discussed in [6]. Service breaks for nurses are finally taken into account as a mandatory rule in [1], [2].

As stated in [4], the HHC service problem is NP-hard. Various solving approaches have been used to solve HHC service problems, including integer linear programming, which is proposed in [3] with the objective is to balance the overall nurse utilization, constraint programming and (meta)heuristics that are introduced in [4], [5], [2] with some variants in the objective function, including minimizing the travel distance of nurses, minimizing the total operational cost, minimizing the unscheduled requests, and remaining the personal relationship between patient and nurse. According to [3], in the literature metaheuristic approaches are usually proposed to solve HHC problems rather than exact approaches, due to the intrinsic complexity of the problem.

In this paper, we focus primarily on scheduling and routing in the context of HHC services. The objective is to devise a model able to provide, once solved, the treatment schedule and the tours of nurses in such a way that a lowest possible operational cost is guaranteed together with a consistent personal relationship between patient and nurse. In addition, in order to obtain a solution that satisfies the reasonable requirements in practice such as the necessity of starting service within a given time window for certain treatments, we take both soft and hard time window into account when designing the model. A maximum allowed value of overtime L is considered to meet the labor regulations regarding working time while reducing as much as possible the number of unfilled requests. Moreover, another important fact of interest addressed in the model is balancing the workload among the nurses. These elements together represent the main novel contribution of the present paper. The rest of the paper is organized as follows. Section II outlines the HHC service problem investigated in the paper, providing a comparison with previous works. Section III presents an integer linear programming formulation of the problem. Section IV presents some computational results stating the obvious applicability and flexibility of the proposed model. Section V concludes the paper.

II. PROBLEM DESCRIPTION

We consider an extended version of the existing HHC service models. The model is defined on a complete directed network containing a single central office called depot. Two main elements of the HHC service are the patients and the nurses. The patients are represented via a set of job J in which a job $j \in J$ is a medical treatment required by patient. Each job $j \in J$ is characterized by a soft time window $[e_j, l_j]$ and a hard time window violations $[lbv_j, ubv_j]$ regarding the starting time of the service at job j. Penalty costs c_{ej} and c_{lj} are paid if the service starts before time e_j or after time l_j, respectively. The service cannot however start before time lbv_j or after time ubv_j. This modeling choice is designed to provide a level of flexibility which is desirable in practice: certain treatments have necessarily to be done within a given time windows, and preferably within a tighter time window. The duration of the service performed at job $j \in J$ is d_j. In additional, each job $j \in J$ requires a certain skill level $\bar{q}_j$ and a set of operation days D_j during the planning horizon which is presented by a set of days D of a week. The traveling time between jobs i and $j \in J$ is finally t_{ij}.

A set of nurses O, representing the nurses available in the team. Each nurse $n \in O$ has a set of skills Q_n representing activities for which she has received a training and is therefore able to cover.

The availability and costs of nurses are modeled as follows. If nurse $n \in O$ is available on day $d \in D$ then $a_n^d = 1$, otherwise $a_n^d = 0$. The cost for each overtime unit C_o is just applied to the official nurses. The contract working time of the official nurses for a planning horizon is WT_n and T is the maximum allowed working time for a day. The maximum allowed value of overtime is L. Nurse $n \in O$ can only work on jobs $j \in J$ that require the skills Q_n.

The objective function of the problem can change from Company to Company, based on the specific needs. We therefore let the planner to decide among the following ones, or a linear combination of them: minimizing the number of unscheduled tasks (these tasks will have to be covered with external nurse not covered by the present model); maximizing the patient-nurse loyalty (patients prefer to be visited every time by the same nurse, ideally); minimizing the total overtime cost, the total cost of time window violation; minimizing the

total waiting time; balance the workload as evenly as possible among the nurses.

III. A MIXED INTEGER LINEAR PROGRAMMING FORMULATION

This section provides a mixed integer linear programming model for the problem described in Section II. The following variables are used by the model:

y_{nj}: if nurse n is assigned to job j $y_{nj} = 1$, otherwise $y_{nj} = 0$

x_{ij}^{nd}: if job j is covered immediately after job i by nurse n on day d $x_{ij}^{nd} = 1$, otherwise $x_{ij}^{nd} = 0$

s_{dn}: the starting time of nurse n on day d

e_{dn}: the ending time of nurse n on day d

st_{dj}: the time at which service begins at job j on day d

o_n: overtime cost paid for nurse n during the planning horizon

twv_{dj}^1: the total cost of time window violation occurring at job j on day d when the service starts before time e_j

twv_{dj}^2: the total cost of time window violation occurring at job j on day d when the service starts after time l_j

f_j: the number of unscheduled tasks of job j

z: defines the minimum value of the number of tasks performed by each nurse for the whole planning horizon

w_n: the total time that nurse n has to spend on waiting for the whole planning horizon

loy_j: the number of different nurses performing job j during the planning horizon

Objective function

Our study aims to obtain a solution in which the patients are offered the best possible achievable quality of service, the utilization of nurses is achieved highest possible and the workload is balance among the nurses within the low operational cost. Therefore, the objective function (1) is a combination of six aims weighted where the weights sum up to 1, and have to be selected by the manager in charge of the scheduling, based on the needs of the Company. These aims are, namely, minimizing the number of unscheduled tasks, remaining a consistent patient-nurse loyalty, minimizing the total of overtime cost, minimizing the total cost of time window violation, minimizing the total waiting time, maximizing the value of the minimum number of tasks for each nurse. Notice that the last objective makes the workload evenly balanced among the nurses.

$$
\begin{aligned}
Min \quad & \alpha_1 \sum_{j \in J, j \neq 0} f_j + \alpha_2 \sum_{j \in J, j \neq 0} loy_j + \alpha_3 \sum_{n \in O} o_n \\
& + \alpha_4 \sum_{d \in D} \sum_{j \in J, j \neq 0} \left(twv_{dj}^1 + twv_{dj}^2 \right) + \alpha_5 \sum_{n \in O} w_n \\
& - \alpha_6 z
\end{aligned} \tag{1}
$$

Subject to

$$loy_j \geq \sum_{n \in O} y_{nj} - 1 \qquad j \in J, j \neq 0 \qquad (2)$$

$$y_{nj} \leq \sum_{d \in D_j} a_n^d \qquad n \in O_j = \{n \in O | q_n = \bar{q}_j\},$$
$$j \in J, j \neq 0 \qquad (3)$$

$$y_{nj} \leq \sum_{d \in D_j} \sum_{i \in J, i \neq j} x_{ij}^{nd} \qquad n \in O_j = \{n \in O | q_n = \bar{q}_j\},$$
$$j \in J, j \neq 0 \qquad (4)$$

$$y_{nj} = 0 \qquad n \in (O \setminus O_j), j \in J, j \neq 0 \qquad (5)$$

$$x_{ij}^{nd} = 0 \qquad n \in O, d \in (D \setminus D_j),$$
$$j \in J, j \neq 0, i \in J, i \neq j \qquad (6)$$

$$x_{ij}^{nd} \leq y_{nj} a_n^d \qquad i, j \in J, i \neq j, j \neq 0, n \in O, d \in D_j \qquad (7)$$

$$f_j = |D_j| - \sum_{d \in D_j} \sum_{n \in O} \sum_{i \in J, i \neq j} x_{ij}^{nd} \qquad j \in J, j \neq 0 \qquad (8)$$

$$o_n \geq C_o \left(\sum_{d \in D} (e_{dn} - s_{dn}) - WT_n \right) \qquad n \in O \qquad (9)$$

$$twv_{dj}^1 \geq c_{ej}(e_j - st_{dj}) \qquad j \in J, j \neq 0, d \in D_j \qquad (10)$$

$$twv_{dj}^2 \geq c_{lj}(st_{dj} - l_j) \qquad j \in J, j \neq 0, d \in D_j \qquad (11)$$

$$\sum_{d \in D} (e_{dn} - s_{dn}) - WT_n \leq L \qquad n \in O \qquad (12)$$

$$z \leq \sum_{d \in D} \sum_{j \in J, j \neq 0} \sum_{i \in J, i \neq j} x_{ij}^{nd} \qquad n \in O \qquad (13)$$

$$w_n = \sum_{d \in D} (e_{dn} - s_{dn}) - \sum_{d \in D} \sum_{j \in J} \sum_{i \in J, i \neq j} x_{ij}^{nd} t_{ij}$$
$$- \sum_{d \in D} \sum_{j \in J, j \neq 0} \sum_{i \in J, i \neq j} x_{ij}^{nd} d_j \qquad n \in O \qquad (14)$$

$$s_{dn} + t_{0j} - M(1 - x_{0j}^{nd}) \leq st_{dj} \qquad n \in O, d \in D_j,$$
$$j \in J, j \neq 0 \qquad (15)$$

$$st_{dj} + d_j + t_{j0} - M(1 - x_{j0}^{nd}) \leq e_{dn} \qquad n \in O, d \in D_j,$$
$$j \in J, j \neq 0 \qquad (16)$$

$$st_{di} + d_i + t_{ij} - M(1 - x_{ij}^{nd}) \leq st_{dj} \qquad d \in D, n \in O,$$
$$i, j \in J \setminus \{0\}, i \neq j \qquad (17)$$

$$e_{dn} - s_{dn} \leq T \qquad n \in O, d \in D \qquad (18)$$

$$\sum_{j \in J, j \neq 0} \sum_{i \in J, i \neq j} x_{ij}^{nd} d_j + \sum_{j \in J} \sum_{i \in J, i \neq j} x_{ij}^{nd} t_{ij} \leq e_{dn} - s_{dn}$$
$$n \in O, d \in D \qquad (19)$$

$$\sum_{n \in O} \sum_{i \in J, i \neq j} x_{ij}^{nd} \leq 1 \qquad j \in J, j \neq 0, d \in D \qquad (20)$$

$$\sum_{j \in J, j \neq 0} x_{0j}^{nd} \leq 1 \qquad n \in O, d \in D \qquad (21)$$

$$\sum_{j \in J, j \neq 0} x_{j0}^{nd} = \sum_{j \in J, j \neq 0} x_{0j}^{nd} \qquad n \in O, d \in D \qquad (22)$$

$$\sum_{i \in J, i \neq j} x_{ij}^{nd} = \sum_{i \in J, i \neq j} x_{ji}^{nd} \qquad n \in O, d \in D_j,$$
$$j \in J, j \neq 0 \qquad (23)$$

$$\sum_{j \in J, j \neq 0} \sum_{i \in J, i \neq j} x_{ij}^{nd} \leq M \sum_{i \in J, i \neq 0} x_{0i}^{nd} \qquad n \in O, d \in D \qquad (24)$$

$$s_{dn} \leq M' \sum_{j \in J, j \neq 0} x_{0j}^{nd} \qquad n \in O, d \in D \qquad (25)$$

$$e_{dn} \leq M' \sum_{j \in J, j \neq 0} x_{0j}^{nd} \qquad n \in O, d \in D \qquad (26)$$

$$lbv_j \leq st_{dj} \qquad d \in D, j \in J, j \neq 0 \qquad (27)$$

$$st_{dj} \leq ubv_j \qquad d \in D, j \in J, j \neq 0 \qquad (28)$$

$$y_{nj} \in \{0, 1\} \qquad n \in O, j \in J \qquad (29)$$

$$x_{ij}^{nd} \in \{0, 1\} \qquad n \in O, d \in D,$$
$$i, j \in J, i \neq j \qquad (30)$$

$$s_{dn}, e_{dn} \geq 0 \qquad n \in O, d \in D \qquad (31)$$

$$twv_{dj}^1, twv_{dj}^2, st_{dj} \geq 0 \qquad d \in D, j \in J \qquad (32)$$

$$o_n, w_n \geq 0 \qquad n \in O \qquad (33)$$

$$f_j, loy_j \geq 0 \qquad j \subset J \qquad (34)$$

$$z \geq 0 \qquad (35)$$

The patient nurse loyalty is presented through the number of different official nurses performing job j during the planning horizon with her high skill. As stated in constraints (2), the loyalty worsens as the value of loy_j increases. Constraints (3) state that nurse n who provides the high skill as the same as the skill required by job j can be assigned to job j only if that nurse is available at least one day among the required operation days of job j. Constraints (4) mean that nurse n cannot be assigned to job j if that nurse never performs job j during the planning horizon. Equations (5) state that nurse n who does not have the high skill required by job j cannot be assigned to job

j. Equations (6) mean that official nurse n will not perform job j on day d which is not the operational day of job j. Constraints (7) mean that official nurse n can not perform job j on day d if she is not available on day d or she has not been assigned to job j even she provides the high skill as the same as the skill required by job j. The number of unscheduled tasks for each job j is given in Equations (8). Equations (9) defines the upper bound of the amount of overtime cost paid for each official nurse n for the whole planning horizon and the upper bounds of the total cost of time window violation occurring at each job j when the service starts before time e_j or after time l_j for each operational day are given in Equations (10) and (11). The constraints (12) guarantee that the total overtime of each nurse cannot exceed the maximum allowed value. Constraints (13) guarantee the workload is balance among the nurses by defining the lower bound of the number of tasks for the whole planning horizon. The total waiting time for the whole planning horizon of each nurse is given in Equations (14). The feasibility of the starting time and the ending time of nurse n on day d are assured in constraints (15) and (16). Constraints (17) guarantee that the starting times of jobs are feasible. Constraints (18) mean that for each nurse, the daily workload does not exceed the maximum allowed working time T. Constraints (19) assure that the tour is long enough so that nurse n can finish up all the tasks on that tour. Equations (20) guarantee that each job is visited exactly once on day d by only one nurse if that job is performed on that day. Constraints (21) and (22) allow the case that on day d nurse n does not need to visit any patient even though that nurse is available on that day. If nurse n performs at least one treatment on day d then she has to start once and end once at the depot. Constraints (23) mean that each nurse n has to leave the patient after visiting. Constraints (24) state that if nurse n does not perform any treatment on day d then that nurse will not visit any patient on that day. If nurse n does not perform any treatment on day d then her starting time and ending time on day d are both equal zero as stated in constraints (25) and (26). The service at job j must be started within its hard time window violation is guarantee in constraints (27) and (28). Constraints (29)-(35) state variables domains.

IV. COMPUTATIONAL RESULTS

In order to demonstrate the applicability for the practical solution of the proposed model as well as its flexibility in that it can be modified for use in multiple different instances of application, we conduct several experiments based on a set of randomly generated home health care scenarios starting from real life data stated in [4]. These real life data sets were provided by German charity and Dutch software company ORTEC in Netherlands. Each scenario is a combination of the number of medical treatments called jobs required by patients (40, 50, 60, 70, 80), the number of nurses (8, 12, 16, 20), six weighted values reflecting the decision on the objective function of the manager regarding the needs of the Company. In particular, weights α_1 and α_2 reflect the purpose of satisfying the demand of patients, weights α_3, α_4, and α_5 present the needs of the Company in terms of economy with the aim of reducing as much as possible operational cost and achieving a highest possible utilization of nurses, weight α_6 states the aim of satisfying the demand of nurses. Table I shows the combination of these six weighted values.

In all scenarios, the planning horizon is a week. There is only one starting point called depot in the system. We assume that all the services provided by the care provider require the skills that belong to the set of five skills regarding medical treatments. The cost of one hour overtime is 15 dollars, the value of four hours is the maximum allowed working time for a day, and the maximum allowed value of overtime for the whole planning horizon is 12 hours. Each job requires a service time of 15 minutes to 60 minutes and a certain skill level from a set of skills. In addition, the number of treatments required by each job is randomly generated up to four treatments during the planning horizon. The traveling times among the locations of jobs are ranged from 20 to 45 minutes. The early starting time randomly generated of each job is any time within a day. Then the late starting time and hard time window violations are generated dependently on the service time and the early starting time. Penalty costs rising when the service starts out of the range of the soft time window are ranged from 10 to 20 dollars for an hour. Each nurse who is available in the team owns at least two certain skill levels from the set of skills and has to be available at least two days during the planning horizon. The contract working time of each nurses is ranged from 56 to 70 hours for a whole week.

All experiments were performed on an Intel(R) Core(TM) i7-3630QM CPU with 2.4GHz and 8.0 GB Ram. The solver is GUROBI 5.5.0 C++ interface with a time limit of three hours for exploring nodes. All results obtained are described in Table II, Table III, and Table IV. The dash symbols under the column "Running time" means that the solver was executed until it reaches the execution time limit. The dash symbols under the column "Gap" means that the optimal solution was found and therefore there is no gap.

Table II summarizes the results of experiments on a set of randomly generated home health care scenarios in which the number of nurses is fixed at the value of 12 and the number of jobs is changed from 40 up to 80. The obtained results summarized in columns 4 and 5 of the table are running times in seconds and percentage gaps between the best value found and the optimal value, respectively. These results reveal that within a time limit of three hours for exploring nodes, even though in the case of the number of job is 70, the optimal solution can be found with the experiment on CoW4 that focuses on only satisfying the demands of patients, the very small gap 0.43% occurs with the experiment on CoW2 that considers the demand of the Company regarding reducing the operational costs to be more important than other demands, or the percentage gaps are always less than 3% with other experiments CoW1 and CoW3, our solver never finds the optimal solutions in the case of the number of jobs is 80 for all instances of the combination of weights. Moreover, there is a significant increase in the percentage gaps when the number of jobs is increased from 70 to 80. When the number of jobs is 40, 50, and 60, the solver quickly found optimal solutions in every the combination of weights in much shorter running times compared to the cases of the number of jobs is 70 and 80. Overall, observations show that with the current resource in which the number of nurses is 6, in order to obtain the best solution values when considering a weighted sum of objectives regarding satisfying simultaneously the demands of patients, nurses, and the Company the Company can operate their home health care system to serve up to 70 jobs. When the aim is to

TABLE I. COMBINATION OF WEIGHTS

Combination of weights	α_1	α_2	α_3	α_4	α_5	α_6
CoW1	0.300	0.300	0.125	0.125	0.125	0.025
CoW2	0.150	0.150	0.250	0.250	0.100	0.100
CoW3	0.200	0.200	0.100	0.100	0.200	0.200
CoW4	0.500	0.500	0.000	0.000	0.000	0.000

only satisfy the demands of patients as the objective function, the number of jobs to be serviced can be 80 jobs.

Regarding changing in resource, we conduct several experiments on a set of randomly generated scenarios in which the number of nurses is changed from 6 up to 20 and the number of jobs is fixed at the value of 80. In Table III, columns 4 and 5 representing running times in seconds and percentage gaps between the best value found and the optimal value, respectively show the summarization of the obtained results. The running time does not change noticeably among these experiments but there is a considerable increase in percentage gap among these experiments. This is not surprising, because by increasing the number of nurses, we also increase the number of variables in the model. As a result of this mechanism, our solver needs more time for exploring the nodes to be able to obtain a best solution that is close to the optimal solution. Therefore, in order to get the best possible solution in a reasonable time, the sufficient value of the number of nurses that the Company should have to operate their home health care service under any the combination of weights in the objective function is six. However, although under the same need in the combination of weights in the overall aim, the difference in the number of nurses causes the difference in the solution value. The better solution value in terms of percentage gaps does not mean that each single aim is better as stated in Table IV presenting the detailed results of a particular scenario in which the number of jobs is 80 and the combination of weights is CoW4. For instance, the number of nurses is 20 causes the worst percentage gap compared to other cases in which the number of nurses is less than 20. However, this case produces the smallest value of the number of unsatisfied requests although the patient nurse loyalty level is much higher than other cases. Therefore, the Company may change their decision on the number of nurses to really meet their needs in a particular circumstance.

V. CONCLUSIONS AND FUTURE WORK

In this paper we address the scheduling and routing issues in home health care services problem in the context of the improvement of life expectancy. We proposed a flexible model able to represent the problem in effective practical terms, taking into account some realistic factors previously not considered in the literature. A mixed integer programming formulation for the problem is also presented. The experiments implemented show obviously that the proposed model is efficient and flexible to procedure practical solutions in the real home health care services problem. Its applicability is useful to help the manager planning their home health care services based on the specific needs of the Company.

The proposed model poses good challenges for future research. In particular, efficient methods to solve the model

TABLE II. EXPERIMENTS ON CHANGING THE NUMBER OF JOBS

Nurses	Jobs	Combination of weights	Running time (seconds)	Gap (%)
6	40	CoW1	489.77	-
6	50	CoW1	2270.33	-
6	60	CoW1	3436.20	-
6	70	CoW1	-	2.57
6	80	CoW1	-	4.84
6	40	CoW2	223.64	-
6	50	CoW2	1183.29	-
6	60	CoW2	1755.12	-
6	70	CoW2	-	0.43
6	80	CoW2	-	6.62
6	40	CoW3	293.45	-
6	50	CoW3	605.07	-
6	60	CoW3	2330.26	-
6	70	CoW3	-	2.70
6	80	CoW3	-	6.65
6	40	CoW4	179.86	-
6	50	CoW4	1080.71	-
6	60	CoW4	2926.53	-
6	70	CoW4	4180.57	-
6	80	CoW4	-	1.31

TABLE III. EXPERIMENTS ON CHANGING THE NUMBER OF NURSES

Nurses	Jobs	Combination of weights	Running time (seconds)	Gap (%)
6	80	CoW1	-	4.84
8	80	CoW1	-	16.53
12	80	CoW1	-	50.81
16	80	CoW1	-	67.96
20	80	CoW1	-	73.18
6	80	CoW2	-	6.62
8	80	CoW2	-	20.87
12	80	CoW2	-	48.89
16	80	CoW2	-	71.60
20	80	CoW2	-	78.40
6	80	CoW3	-	6.65
8	80	CoW3	-	19.19
12	80	CoW3	-	51.70
16	80	CoW3	-	69.65
20	80	CoW3	-	78.36
6	80	CoW4	-	1.31
8	80	CoW4	-	19.55
12	80	CoW4	-	33.33
16	80	CoW4	-	62.20
20	80	CoW4	-	65.63

TABLE IV. DETAILED RESULTS OF A PARTICULAR SCENARIO IN WHICH THE NUMBER OF JOBS IS 80 AND THE COMBINATION OF WEIGHTS IS COW4

Nurses	Running time (seconds)	Gap (%)	Patient-Nurse loyalty level	Unscheduled tasks total
6	-	1.31	0	153
8	-	19.55	0	133
12	-	33.33	0	102
16	-	62.20	2	80
20	-	65.63	3	61

have to be devised: it has to be expected that for large instances commercial solvers will be unable to solve the mixed integer programming formulation we proposed directly. Therefore heuristic methods, possibly based on mathematical programming will have to be developed. Extensions to the model presented are possible to make it even more realistic, possibly without increasing too much the complexity. For example, together with the official nurses who are available in the team, it is possible to define external nurses who are available on the market to operate in case the official nurses are not enough to cover the given demand, but with a very high cost. Moreover, it is possible to associate to each nurse both a set of high skills and a set of low skills, representing activities for which she has a full training or a partial training (but sufficient to operate), respectively. Nurses are able to operate on their low skills in case of emergency, but normally should operate on their high skills only. Notice that nurses working on their low skills can develop experience in these skills by practical experience, and in the long run these skills can be added to their high skills, with the associated benefit for the nurses' career (similar ideas were proposed in [7]). Finally, uncertainty about the effective presence of the nurses (they might call sick) and about travel and service time could be inserted into the model.

ACKNOWLEDGMENT

Thi Viet Ly Nguyen was supported by the Swiss National Science Foundation through project 200020-140315/1: "Robust Optimization II: applications and theory"

REFERENCES

[1] A. Trautsamwieser and P. Hirsch, "Optimization of daily scheduling for home health care services," *Journal of Applied Operational Research*, vol. 3, no. 3, pp. 124–136, 2011.

[2] A. A. Cire and J. N. Hooker, "A heuristic logic-based benders method for the home health care problem," 2012.

[3] P. Cappanera and M. G. Scutella, "Joint assignment, scheduling and routing models to home care optimization: a pattern based approach," 2013.

[4] S. Nickel, M. Schröder, and J. Steeg, *Planning for home health care services.* Fraunhofer ITWM, 2009.

[5] T. Macdonald, "Metaheuristics for the consistent nurse scheduling and routing problem."

[6] P. Cappanera and M. G. Scutella, "Home care optimization: impact of pattern generation policies on scheduling and routing decisions."

[7] E. J. Nilssen, M. Stølevik, E. L. Johnsen, and T. E. Nordlander, "Multi-skill shift design for norwegian hospitals," *Journal of Applied Operational Research*, vol. 3, no. 3, pp. 137–147, 2011.

Simulation methods in real option valuation

Tero Haahtela

Aalto University, BIT Research Centre
Helsinki, Finland
tero.haahtela@aalto.fi

Abstract— **This paper discusses different simulation based methods applied for valuing real options. The methods are classified as classical simulation, cash flow simulation for pseudo underlying asset simulation, and advanced simulation based optimization methods for valuing real options. These different classes and their evolution are presented. The purpose is to illustrate the variety of alternative methods and how they have made it easier to value certain kind of investments with real options, and even made it possible to solve more complex real option investment problems than before.**

Keywords—real option valuation, simulation

I.   INTRODUCTION

Simulation is the imitation of a real-world process or system with a mathematical model. In simulation, a computer is used to evaluate a model numerically, and data is gathered to estimate the true characteristics of the model describing the system [1]. The main characteristic that makes simulation attractive to real option valuation is its ability to cope with uncertainty in a very simple way [2]. When compared with a purely mathematical analysis, simulation allows models that are closer to the real world with more complexity and fewer restricting assumptions of reality [3]. Monte Carlo methods are usually intuitive and mathematically less demanding than highly sophisticated models of equations. By simulation, mathematical complexity is changed into a numerical problem, and work is transferred to computer [4]. Interactions and sensitivities of the system to various factors can be examined by changing parameter values (ibid.). As a result, simulation is a flexible way to analyze and value business situations with different alternatives [5].

Simulation approximates numerically the correct rather than exact solutions. The accuracy of results increases with the larger number of iterations. For financial options, both precision and efficiency and their optimal balance are relevant as the prices fluctuate continuously and the portfolios need to be re-balanced based on real-time information. Therefore, analytical solutions or approximations of analytical solutions are preferred over numerical simulation with financial options. In case of real option problems, computationally less efficient methods are usually sufficient for given level of precision. Because there is no need to continuous re-balancing, also computationally less efficient methods may be used. Understandability, robustness and flexibility are main concerns in selection of valuation approach. Another aspect is that the required level of precision in case of financial options is far more crucial than the required accuracy in case of real options.

Precision is not that essential in real option valuation in comparison with capability to approximate and model the reality on a sufficient level [6]. Most practitioners prefer an approximate solution to a realistic problem over an exact solution to an oversimplified problem that is only a rough approximation of reality. As a result, simulation based alternatives are much better suitable for real options than financial options.

II.   CLASSICAL SIMULATION METHODS FOR REAL OPTIONS

Classical simulation methods refer to the older simulation based option valuation methods. They were originally developed for valuing financial options, but later they have been applied to real options as well. Typically these approaches are easier to use and they do not need special software to be calculated.

A. Simulation for numerical integration

Monte Carlo method was invented in the late 1940s by Stanislaw Ulam and Von Neumann while working on nuclear weapon projects. Original use of Monte Carlo method was Monte Carlo integration. Instead of attempting to carry out directly high-dimensional integrations involving the transition probabilities of many possible intermediate events and states by the use of lattice methods and regular grids, MC single chains of events are sampled at randomly chosen points in MC. While this approach could be used for valuing financial options and real options as well, increase in computing power and more advanced valuation and simulation methods were invented before the real option valuation was introduced.

B. Path sampling for an empirical distribution

Direct Monte Carlo simulation method (also called trace driven simulation) suggested by Boyle [7] is based on the idea of approximating the stochastic process numerically. The valuation procedure consists of the following steps:

1) Random path for the underlying asset is sampled in a risk-neutral world and the payoff of the option at the maturity is calculated.

2) This is repeated at least several thousand times

3) The mean of the sample payoffs is calculated to get an estimate of the expected payoff in a risk-neutral world

4) Payoff is discounted at risk-free rate to get the present option value.

Direct Monte Carlo simulation has several advantages. It can accommodate complex payoffs and complex stochastic processes and take into account situations where payoff depends on some function of the whole path followed by a variable, not just its terminal value. Payoffs may occur at several times during the life of the derivative [8]. The approach has become more attractive relative to other numerical techniques because it is flexible as well as easy to implement and modify [9]. The terminal value distribution may also be of any arbitrary shape or form and it does not need to be related to any stochastic process. In the most straightforward way, the results can be applied as: value of the European option is defined as *Risk Adjusted Success Probability * (Benefits – Costs)* discounted to present value [10].

C. Sampling for a theoretical distribution

Instead of having to sample the whole paths, sometimes we only need to know or sample the terminal value distribution values. While there are closed-form solutions available for partial differential equations for many option types (normal distribution, displaced lognormal distribution), there also distribution types that do not have this property. MCS is not dependent of distribution form, and therefore sampling for a theoretical distribution can solve any European option this way.

D. Simulation for valuing path-dependent options

One significant strength of Monte Carlo simulation is capability to value path dependent options whose payoff depends on the path followed by the price of the underlying asset, and not just on its final value. Examples of such options are Asian options whose value depends on the average price of the underlying asset. The price of such option is mean of the sample values, which are calculated by sampling a random path for the underlying asset in a risk-neutral world, calculating the path-dependent payoff, and discounting the payoff at the risk-free interest rate [8]. The only difference is that in a single simulation run the payoff is the sum of the discretized path values instead of being only dependent of the final payoff value.

The approach is efficient for real options purposes, because it allows use of any kind of stochastic processes including jumps, mean-reversion, seasonalities, and other factors. Typical areas of application for these options are projects related to energy derivatives. Because some of these options are rather operational by nature, their stochastic processes may also differ from those typical for financial derivatives modeling. While financial models often assume geometric Brownian motion and perhaps mean-reversion and jumps, industrial operational processes may also have trends, seasonality, cyclicality, S-shape curves and even extreme value like events. Haahtela [11] generally describes these kinds of options as cash flow simulation embedded options as their value is based on choosing the optimal forward-looking decision in each time step during a single cash flow calculation simulation run.

E. American options with least squares regression

Monte Carlo simulation is by nature a forward-looking approach, and therefore it was long considered not to be able handle early exercise of American type options. However, Longstaff and Schwartz [12] popularized and improved the Carriere's suggestion [13] how to value American options with simulation by applying least squares regression. The method uses least-squares analysis to determine the best-fit relationship between the value of continuing and values of relevant variables at each time an early exercise decision has to be made. At each exercise time, the payoff from immediate exercise is compared with the expected payoff from continuation. The optimal exercise strategy is determined by the conditional expectation function that is estimated from the cross-sectional information in the simulation by using least squares. This is found by regressing the ex post realized payoffs from continuation on functions of the values of the state variables, and this fitted value provides a direct estimate of the conditional expectation function. A complete specification of the optimal exercise strategy along each path is obtained by estimating the conditional expectation function for each state.

The method itself is of use in valuing American options, but it has also been a key idea for several cash flow simulation based volatility estimations methods.

III. CASHFLOW SIMULATION FOR CREATING A PSEUDO UNDERLYING ASSET

One way to apply Monte Carlo simulation for ROV is to simulate cash flow calculation and its uncertainties and then use this information for estimating the underlying asset price and its stochastic behavior. Monte Carlo simulation on cash flows consolidates a high-dimensional stochastic process of several correlated variables into a low-dimension stochastic summary process. Usually this process is a univariate geometric Brownian motion, in which case the volatility parameter σ of the underlying asset is estimated by calculating the standard deviation of the simulated probability distribution for the rate of return.

A. Marketed Asset Disclaimer

Most common approach of this category is Copeland and Antikarov's [14] Marketed Asset Disclaimer (MAD) approach suggesting that the present value of the project cash flows without options is the best unbiased estimator of the market value of the project were it a traded asset. According to this Marketed Asset Disclaimer, or MAD assumption, the project itself is the underlying asset of the replicating portfolio. Another assumption, based on Samuelson's proof [15], is that the variations in the value of the project follow a random walk, thus justifying use of geometric Brownian motion as the stochastic process for the underlying asset. Along with the MAD assumption, Copeland and Antikarov also suggest a four-step valuation process framework with a specific volatility estimation procedure for valuing investments with real options.

B. Cash flow simulation based volatility estimation

Several authors have corrected and extended the original logarithmic volatility estimation procedure suggested in [14]. The methods are mostly based on the MAD assumption and some modification on the least squares approach suggested by

Longstaff and Schwartz [12]. Godinho [16] corrected the original volatility estimation procedure, Haahtela [17] and Brandao, Dyer and Hahn [18] extend volatility to be stochastic. Haahtela [19] presents a practical approach for parameterizing a more general displaced diffusion process with time-varying volatility. With a spreadsheet program and a Monte Carlo simulation software, the present value of the cash flows and the cash flow state variable values are recorded for each time period during cash flow simulation runs. Then a regression analysis is run for relating the PV for each time period to the corresponding cash flow state variables. Each regression equation provides an estimate of the expected present value as a function conditioned on the resolution of all uncertainties up to that time. Then, the basic regression statistics of Pearson's correlation R^2 and sum of squares error (or standard error) for each equation provide all the information required for estimating how the standard deviation and displaced volatility change over time. Brandao, Dyer and Hahn [18] use similar approach for volatility estimation and call it as generalized conditional expectations approach. They also show that the approach is suitable for cases with a mean-reverting process and stochastic jump to the variable cost process.

C. Non-parametric real option valuation based on cash flow simulation

Based on the same MAD assumptions, Wang and Dyer [20] suggest a non-parametric implied binomial tree based on simulated cash flows. The approach implements multifactor real option valuation techniques using risk neutral probabilities inferred from the simulated market information. Their algorithm relaxes the distribution assumption for the underlying project value and provides an alternative approach to valuing many high-dimensional real options problems with generalized diffusion processes.

IV. ADVANCED SIMULATION BASED OPTIMIZATION METHODS FOR REAL OPTION VALUATION

Earlier mentioned simulation methods are sufficiently straightforward to implement even with ordinary spreadsheet program (and with a little help of a simulation software). However, there are many complex, non-linear, and path-dependent cases with possible delays and feedback loops. System dynamics, neural networks and genetic algorithms are each simulation based approach families with many different variations. They have been suggested to solve some of these issues in real option valuation. While the methods may be more capable approaches, their use requires special software and knowledge on their properties and capability to choose the right model and algorithm for each case.

A. System dynamic models in ROV

Some investment problems may be so complex that the previously mentioned path-dependent approaches might not find the optimal solution. Some projects are characterized by stochastic processes embedded in non-linear feedback structures with delays. System dynamics models may be used to estimate the cash flow resulting from these projects. If these projects include managerial flexibility, a correct financial evaluation of these cash flows requires the use of real options methodology. Ford and Sobek [21] apply these approaches together for switching among alternatives in product development, while Tan, Anderson, Dyer and Parker [22] demonstrate how to apply this approach in valuing wind power industry.

B. Neural networks based optimization

Taudes et al. [23] and [24] suggest using neural networks to value complex real option cases when analytical solutions do not exist. The basic idea is to approximate the value function of a dynamic program by a neural net, where the election of the network weights is done via simulated annealing. Common procedure is to use multi-layer feed forward network. In the first stage, samples of training data are generated from simulation or measurement. In the second stage, neural network is trained by adjusting its weights so that the network predicts best output matches, usually by minimizing some norm of the error function between the predicted output of the neural network and the targeted outputs. The main benefits of this method as compared to traditional approximation techniques are that i) there are no restrictions on the type of the underlying stochastic process, and ii) no limitations on the set of possible actions. This makes the approach especially attractive for valuing real options in flexible investments. The approach has been applied for valuing flexibility for costly switching in production between several products under various conditions [23].

C. Genetic algorithms

Genetic algorithms are search heuristics that mimic the process of natural evolution. The approaches are based on selecting and recombining iteratively the most potential alternatives to generate a new set of solutions. Gradually the solution evolves towards the optimum. The method is common for finding solutions to complex optimization problems.

While Monte Carlo simulation is considered a good way to model several uncertainties and their complex relationships, it is not very good for finding optimal solutions for such optimization problems. Dias and Rivera [25] present a model of dynamic hedging aimed at optimization and using genetic algorithms and Monte Carlo simulation. The approach offers new possibilities for energy investment applications of dynamic hedging based on real options theory. The case example showed that several competing threshold alternative curves drove the search to the "optima region," within which any one solution is indistinguishably close to an absolute-optimum solution [25]. This result provided new insights about real options evaluation not previously addressed in the literature.

V. SUMMARY AND CONCLUSIONS

This paper presented and classified several different approaches for valuing real options with simulation. As the examples illustrate, there are many different ways to utilize simulations in real option valuation. Simulation based approaches are also likely to gain more popularity among practitioners and decision makers as they learn about the latest

Proceedings of FORS40 - 2013 Lappeenranta, Finland - ISBN 978-952-265-435-9

discoveries in the field and fully understand the flexibility of the simulation based models. Advanced simulation methods have made it possible to solve more complex investment problems than before. The main advantage, however, is that simulation offers an easier way to model and understand many investment cases with added realism and less restricting assumptions than most alternative real option valuation approaches.

VI. REFERENCES

[1] A. Law and D. Kelton, *"Simulation modeling and analysis"*. 3rd ed. McGraw & Hill, USA, 2000.

[2] G. Cortazar, "The Valuation of Natural Resources", in *Real Options and Business Strategy – Applications to decision Making*, (Ed. L. Trigeorgis), Risk Books, pp. 263-278, 2000.

[3] P. Jäckel, *"Monte Carlo methods in finance"*, John Wiley & Sons, 2002.

[4] D. McLeish, Monte Carlo Simulation and Finance, Wiley Finance, 2005.

[5] R. Razgaitis, "Dealmaking using real options and monte carlo analysis", Wiley Finance, 2003.

[6] M. Amram and N. Kulatilaka, "Real Options: Managing Investment in an Uncertain World", Harvard Business School Press, 1999.

[7] P. Boyle, "Options: a Monte Carlo approach", *Journal of Financial Economics*, Vol. 4, pp. 323–338, 1977.

[8] J. Hull, *"Options, futures and other derivatives"*, 6th edition. Prentice-Hall, 2006.

[9] P. Boyle, M. Broadie and P. Glasserman, "Monte Carlo methods for security pricing", *Journal of Economic Dynamics and Control*, Vol.21 , pp. 1267-1321, 1997.

[10] S. Mathews, V. Datar and B. Johnson, "A Practical Method for Valuing Real Options: The Boeing Approach," *Journal of Applied Corporate Finance*, vol. 19, pp. 95-104, 2007.

[11] T. Haahtela, Cash flow simulation embedded real options, ICAOR 2010, Lecture Notes on Management Science, pp. 418-430, 2010.

[12] F. Longstaff and E. Schwartz, "Valuing American options by simulation: A simple least-squares approach", *Review of Financial Studies*, Vol. 14, No. 1, pp. 113–147, 2001.

[13] J. Carriere, "Valuation of early-exercise price of options using simulations and nonparametric regression", *Insurance: Mathematics and Economics*. Vol. 19, pp. 19-30, 1996.

[14] T. Copeland and V. Antikarov, *Real Options: a practitioner's guide*. New York, NY: Texere, 2001.

[15] P. Samuelson, "Proof That Properly Anticipated Prices Fluctuate Randomly," Industrial Management Review, vol. 6, pp. 44-49, 1965.

[16] P. Godinho, "Monte Carlo Estimation of Project Volatility for Real Options Analysis", *Journal of Applied Finance*, Vol. 16, No. 1, Spring/Summer 2006.

[17] T. Haahtela, "Recombining Trinomial Tree for Real Option Valuation with Changing Volatility". 14th Annual International Conference on Real Options, 16-19 June 2010, Rome, Italy. Available at SSRN: http://ssrn.com/abstract=1932411 2010.

[18] L. Brandao, J. Dyer and W. Hahn, "Volatility estimation for stochastic project value models", *European Journal of Operational Research*, Vol. 220, pp. 642-648, 2012.

[19] T. Haahtela, "Estimating Changing Volatility in Cash Flow Simulation-Based Real Option Valuation with the Regression Sum of Squares Error Method", Journal of Real Options, Vol. 1 No. 1, pp. 33-52, 2011.

[20] T. Wang and J. Dyer, "Valuing Multifactor Real Options Using an Implied Binomial Tree", *Decision Analysis*, Vol. 7, No. 2, pp. 185—195, 2010.

[21] D. Ford and D. Sobek, "Adapting real options to new product development by modeling the second Toyota Paradox", *IEEE Transactions on Engineering Management*. Vol. 52, No. 2, pp. 1-11. May, 2005.

[22] B. Tan, E. Anderson, J. Dyer, G. Parker, "Using Binomial Decision Trees and Real Options Theory to Evaluate System Dynamics Models of Risky Projects", Proceedings of the 27th International Conference of the System Dynamics Society, 26-30 July, Albuquerque, USA, 2009.

[23] A. Taudes, M. Natter and M. Trcka, "Real option valuation with neural networks", *Intelligent Systems in Accounting, Finance and Management*, Vol. 7, pp. 43-52, 1998.

[24] C. Charalambous and S. Martzoukos, "Hybrid artificial neural networks for efficient valuation of real options and financial derivatives", *Computer Management Science* 2, pp. 155-161, 2005.

[25] M. Dias and M. Rivera, "Real Options Valuation in Energy Investment Projects: Modeling Hedging Strategies Using Genetic Algorithm Software", Problems and Perspectives in Management, pp. 234-247, 2004.

A Practical Analytical Model for Valuing Early Stage Investments Using Real Options

Yuri Lawryshyn

Abstract—In this work, we build on a previous real options approach that utilizes managerial cash-flow estimates to value early stage project investments. Through a simplifying assumption, where we assume that the managerial cash-flow estimates are normally distributed, we derive a closed-form solution to the real option problem. The model is developed through the introduction of a market sector indicator, which is assumed to be correlated to a tradeable market index and drives the project's cash-flow estimates. In this way we can model a cash-flow process that is partially correlated to a traded market index. This provides the mechanism for valuing real options of the cash-flow in a financially consistent manner under the risk-neutral minimum martingale measure. The method requires minimal subjective input of model parameters and is very easy to implement.

Keywords—Real Options; Managerial Estimates; Closed-Form Solution; Cash-Flow Replication; Project Valuation.

I. INTRODUCTION

Real option analysis (ROA) is recognized as a superior method to quantify the value of real-world investment opportunities where managerial flexibility can influence their worth, as compared to standard net present value (NPV) and discounted cash-flow (DCF) analysis. ROA stems from the work of Black and Scholes (1973) on financial option valuation. Myers (1977) recognized that both financial options and project decisions are exercised after uncertainties are resolved. Early techniques therefore applied the Black-Scholes equation directly to value put and call options on tangible assets (see, for example, Brennan and Schwartz (1985)). Since then, ROA has gained significant attention in academic and business publications, as well as textbooks (Copeland and Tufano (2004), Trigeorgis (1996)).

While a number of practical and theoretical approaches for real option valuation have been proposed in the literature, industry's adoption of real option valuation is limited, primarily due to the inherent complexity of the models (Block (2007)). A number of leading practical approaches, some of which have been embraced by industry, lack financial rigor, while many theoretical approaches are not practically implementable. The work presented in this paper is an extension of an earlier real options method co-developed by the author (see Jaimungal and Lawryshyn (2010)). In our previous work we assumed that future cash-flow estimates are provided by the manager in the form of a probability density function (PDF) at each time period. As was discussed, the PDF can simply be triangular

Yuri Lawryshyn is an Associate Professor with the Centre for the Management of Technology and Entrepreneurship at the Department of Chemical Engineering and Applied Chemistry, University of Toronto, e-mail: yuri.lawryshyn@utoronto.ca.

(representing typical, optimistic and pessimistic scenarios), normal, log-normal, or any other continuous density. Second, we assumed that there exists a *market sector indicator* that uniquely determines the cash-flow for each time period and that this indicator is a Markov process. The market sector indicator can be thought of as market size or other such value. Third, we assumed that there exists a tradable asset whose returns are correlated to the market sector indicator. While this assumption may seem somewhat restrictive, it is likely that in many market sectors it is possible to identify some form of market sector indicator for which historical data exists and whose correlation to a traded asset/index could readily be determined. One of the key ingredients of our original approach is that the process for the market sector indicator determines the managerial estimated cash-flows, thus ensuring that the cash-flows from one time period to the next are consistently correlated. A second key ingredient is that an appropriate risk-neutral measure is introduced through the minimal martingale measure[1] (MMM) (Föllmer and Schweizer (1991)), thus ensuring consistency with financial theory in dealing with market and private risk, and eliminating the need for subjective estimates of the appropriate discount factor typically required in a DCF calculation.

In this work, we show how our method can lead to an analytical solution, if it is assumed that the possible cash-flows are normally distributed. While this assumption may seem somewhat restrictive, very often, when managers estimate cash-flows (or revenues or expenses) the estimate, itself, is uncertain enough that assumptions regarding the distribution type ultimately only add "noise" to already "noisy" assumptions.

II. REAL OPTIONS IN PRACTICE

A. Project Valuation in Practice

According to Ryan and Ryan (2002), 96% of Fortune 1000 companies surveyed indicated that NPV/DCF was the preferred tool for valuing capital budget decisions, and 83% chose the WACC (weighted average cost of capital) as the appropriate discount factor. While the methodology is well understood, we review a few important details to highlight the main assumptions regarding the DCF approach that are often glossed over or not fully appreciated by practitioners. As discussed in most elementary corporate finance texts (e.g. Berk and Stangeland (2010)) the WACC (equivalently the return on assets) for a company does not change with the debt to

---

[1]As we discuss in Section III, the risk-neutral MMM is a particular risk-neutral measure which produces variance minimizing hedges.

equity ratio when there are no tax advantages for carrying debt. Therefore, without loss of generality, in this discussion, we will assume that the company is fully financed through equity and thus, the cost of equity, r_E, will be the WACC.

The standard approach to estimating r_E is through the Capital Asset Pricing Model (CAPM), which has a significant list of assumptions (again, the reader is referred to any standard text). The CAPM expected return on equity for a given i-th company is estimated as follows,

$$\mathbb{E}[r_{E_i}] = r_f + \beta_i(\mathbb{E}[r_{MP}] - r) \tag{1}$$

where r is the risk-free rate, r_{MP} is the return on the market portfolio and

$$\beta_i = \frac{\rho_{i,MP}\sigma_i}{\sigma_{MP}} \tag{2}$$

where $\rho_{i,MP}$ is the correlation of the returns between the i-th company and the market portfolio, σ_i is the standard deviation of the returns of the i-th company, and σ_{MP} is the standard deviation of the returns of the market portfolio. The point of introducing this well known result is to highlight the fact that by applying the WACC for the valuation of a project, by definition, the manager is assuming that the project has a similar risk profile, compared to the market, as does the company, i.e. that the ρ and σ for the project are identical to that of the company. Clearly, this will often not be the case and a simple thought experiment can illustrate the point.

Consider the case where a manager needs to make two decision: the first is whether to buy higher quality furniture with a longer lifetime versus lesser quality with a shorter lifetime; the second is whether to invest in the development of an enhancement to an existing product line or the development of a new product. The standard DCF approach would use the WACC to discount each of the alternative projected future cash-flows. Clearly, this is wrong. The *value* of the furniture cash-flows likely has little volatility and might, only slightly, be correlated to the market. The product enhancement is likely more closely aligned with the risk profile of the company, while the new product development may, or may not be, highly correlated to the market, but will likely have a high volatility. Our proposed approach decouples the ρ and σ for each project being valued. For the case of valuing the project as of time 0, or, as will be discussed further, for the case of a European real option, the effective σ of the project is determined by the uncertainty in the managerial cash-flow estimates. The ρ for the project does require estimation, however, we provide a mechanism to estimate it as well.

B. Real Options in Practice

As has been well documented (Dixit and Pindyck (1994), Trigeorgis (1996), Copeland and Antikarov (2001)), the DCF approach assumes that all future cash-flows are static, and no provision for the value associated with managerial flexibility is made. To account for the value in this flexibility numerous practical real options approaches have been proposed. Borison (2005) categorizes the five main approaches used in practice, namely, the Classical, the Subjective, the Market Asset Disclaimer (MAD), the Revised Classical and the Integrated. As

discussed by Borison (2005), each method has its strengths and weaknesses. As Borison points out, all of the five approaches at that time, had issues with respect to implementation.

Arguably, one of the promising approaches is the MAD method of Copeland and Antikarov (2001). A number of practitioners have utilized the MAD method (see, for example, Pendharkar (2010)), however, the theoretical basis for the model is debated. Brando, Dyer, and Hahn (2012) have highlighted issues with the way the volatility term is treated in the MAD approach and have recommended an adjustment.

Both the Revised Classical (see Dixit and Pindyck (1994)) and Integrated (see Smith and Nau (1995)) approaches recognize that most projects consist of a combination of market (systematic / exogenous) and private (idiosyncratic / endogenous) risk factors. The latter provides a mechanism to value the market risk of a project through hedging with appropriate tradeable assets while private risk is valued by discounting expected values at the risk-free rate. Properly applied, the integrated approach is consistent with financial theory. As Borison (2005) notes, the integrated approach requires more work and is more difficult to explain, but is the only approach that accounts for the fact that most corporate investments have both market and private risk. As we will show, our approach also accounts for both market and private risks, yet is easy to implement, and, in its simplest form, easy to understand as well.

III. MODEL DEVELOPMENT

A key assumption of the MAD approach, as well as the methods proposed by others, such as Datar and Mathews (2004) and Collan, Fullér, and Mezei (2009), is that the risk profile of the project is reflected in the distribution of uncertainty provided by managerial cash-flow estimates. In Jaimungal and Lawryshyn (2010), we introduced a "Matching Method", where we assumed that there exists a *market sector indicator* Markov process that ultimately drives the managerial-supplied cash-flow estimates. In Jaimungal and Lawryshyn (2011), we extended our matching approach for the more practical case where managers provide uncertain revenue and gross margin percent (GM%) estimates. While both models provide a mechanism to value cash-flows and real options using a simple spreadsheet, because we assumed that the cash-flows, or revenues and GM%, estimates were triangularly distributed, numerical methods were required to solve the project cash-flow and real option valuations. In our work with managers in implementing our models, we realized that very often managers prefer to provide simple estimates consisting of a particular value with a given level of uncertainty. For example, often, managers are just as comfortable to provide a revenue estimate of, say $100 \pm $20, as they are to provide low, medium and high estimates, such as $70, $100, and $120. In the former case, we can assume that the distribution of the estimate is normal. This assumption leads to an analytical solution for our methodology.

In the following subsections we develop our methodology for the case of normally distributed managerial estimates. First, we briefly review our Matching Method for the general case, where we derive a function, φ, that links the market sector indicator process to a normally distributed managerial cash-flow estimate. Then we utilize this function in the risk-neutral minimum martingale measure to develop a very simple expression for the valuation of cash-flows. Next, we develop an analytical formula to value a real option based on the cash-flow estimates, which is essentially a simplification of the model presented in Jaimungal and Lawryshyn (2010).

A. Matching Managerial Estimates

Our objective is to provide a consistent dynamic model that can replicate cash-flow distribution estimates provided by managers. By applying the minimum martingale measure, we are assured of a financially consistent valuation. As such, we assume that there exists a traded market index I_t which the manager can invest in, and we assume that the price of the index follows geometric Brownian motion (GBM),

$$\frac{dI_t}{I_t} = \mu dt + \sigma dB_t, \tag{3}$$

where B_t is a standard Brownian motion under the real-world measure $\mathbb{P}$. Here, we will assume that the manager has provided cash-flow estimates at times T_k, where $k = 1, 2, ..., n$. Specifically, we assume the manager has supplied a cash-flow estimate of the form $\mu_k^F \pm 2\sigma_k^F$ and we assume the resulting distribution to be normal, i.e., $N(\mu_k^F, (\sigma_k^F)^2)$. We assume the cash-flow process to be F_t.

We introduce an observable, but not tradable, *market sector indicator* process A_t which we use to drive the cash-flow estimates provided by the manager. We assume the process to be a standard Brownian motion (BM)

$$dA_t = \rho_{FI}dB_t + \sqrt{1 - \rho_{FI}^2}dW_t^F, \tag{4}$$

where dW_t^F is a standard Brownian motion under the real-world measure $\mathbb{P}$ independent from B_t, and ρ_{FI} is a constant $(-1 \leq \rho_{FI} \leq 1)$. As discussed in Jaimungal and Lawryshyn (2011), choosing a BM instead of a GBM will have no bearing on the results of the valuation from a European real option perspective.

Next, we introduce the collection of functions $\varphi_k^F(A_t)$ such that at each T_k, $F_k = \varphi_k^F(A_{T_k})$. Furthermore, at each cash-flow date T_k we match the the distribution F_k to the cash-flow distribution supplied by the manager, namely, $\Phi\left(\frac{a - \mu_k^F}{\sigma_k^F}\right)$, where $\Phi(\bullet)$ is the standard normal distribution. Thus, we require

$$\mathbb{P}(F_T < a) = \Phi\left(\frac{a - \mu_k^F}{\sigma_k^F}\right). \tag{5}$$

Proposition 1. The Replicating Cash-Flow Payoff. *The cash-flow payoff function $\varphi_k^F(a)$ which produces the managerial specified distribution $\Phi\left(\frac{a - \mu_k^F}{\sigma_k^F}\right)$ for the cash-flows at time T_k, when the underlying driving uncertainty A_t is a BM, and $A_0 = 0$, is given by*

$$\varphi_k^F(a) = \frac{\sigma_k^F}{\sqrt{T_k}}a + \mu_k^F. \tag{6}$$

Proof. We seek $\varphi(.)$ such that $\mathbb{P}(\varphi(A_T) \leq s|\mathcal{F}_0) = \Phi\left(\frac{s-\mu}{\sigma}\right)$. Since,

$$A_{T|\mathcal{F}_0} \overset{d}{=} \sqrt{T}Z \quad \text{where} \quad Z \underset{\mathbb{P}}{\sim} \mathcal{N}(0,1),$$

we have that

$$\mathbb{P}(\varphi(A_T) \leq s|\mathcal{F}_0) = \mathbb{P}\left(\sqrt{T}Z \leq \varphi^{-1}(s)\right)$$
$$= \Phi\left(\frac{\varphi^{-1}(s)}{\sqrt{T}}\right) \triangleq \Phi\left(\frac{s-\mu}{\sigma}\right).$$

Consequently,

$$\varphi(A_T) = \frac{\sigma}{\sqrt{T}}A_T + \mu$$

and the proof is complete. $\qquad\square$

We now have a very simple expression for φ which makes the valuation of risky cash-flows very simple.

B. Valuation of Risky Cash-Flows

As discussed in Jaimungal and Lawryshyn (2010), the real-world pricing measure should not be used, and instead, we propose the risk-neutral measure $\mathbb{Q}$, corresponding to a variance minimizing hedge. Under this measure, we have the following dynamics

$$\frac{dI_t}{I_t} = r\, dt + \sigma\, d\widehat{B}_t, \tag{7}$$

$$dA_t = \widehat{\kappa}\, dt + \rho_{FI}\, d\widehat{B}_t + \sqrt{1 - \rho_{FI}^2}\, d\widehat{W}_t^F, \tag{8}$$

where $\widehat{B}_t$ and $\widehat{W}_t^F$ are standard uncorrelated Brownian motions under the risk-neutral measure $\mathbb{Q}$ and the risk-neutral drift of the indicator is

$$\widehat{\kappa} = -\rho_{FI}\frac{\mu - r}{\sigma}. \tag{9}$$

We emphasize that the drift of the indicator is precisely the CAPM drift of an asset correlated to the market index and is a reflection of a deeper connection between the MMM and the CAPM as demonstrated in (Cerny 1999). Given this connection, our reliance on parameter estimation is similar to those invoked by standard DCF analysis when the WACC is used to discount cash-flows and the cost of equity is estimated using CAPM. In DCF analysis the CAPM drift is, however, estimated based on the company's beta, while in our approach, the CAPM drift derives from historical estimates of the sector indicator and traded index dynamics. Furthermore, the *riskiness* of the project is appropriately captured by the distribution of the cash-flows. Consequently, our approach is more robust. Given the risk measure $\mathbb{Q}$, the values of the cash-flows can now be computed.

Proposition 2. Value of Risky Cash-Flow Estimates. *For a given set of cash-flow estimates, normally distributed with mean μ_k^F and standard deviation σ_k^F, given at times T_k, where $k = 1, 2, ..., n$, the value of these cash-flows at time $t < T_1$ is given by*

$$V_t(A_t) = \sum_{k=1}^{n} e^{-r(T_k - t)}\left(\frac{\sigma_k^F}{\sqrt{T_k}}\left(A_t + \widehat{\kappa}(T_k - t)\right) + \mu_k^F\right),$$
$$\tag{10}$$

and for the case where $t = 0$,

$$V_0 = \sum_{k=1}^{n} e^{-rT_k} \left(\widehat{\kappa} \sigma_k^F \sqrt{T_k} + \mu_k^F \right). \tag{11}$$

Proof. We begin by noting that from equation (8) we can write

$$A_{T|\mathcal{F}_t} \stackrel{d}{=} A_t + \widehat{\kappa}(T - t) + \sqrt{T} Z \tag{12}$$

where $Z \underset{\mathbb{Q}}{\sim} \mathcal{N}(0, 1)$. The value of the cash-flows are given as

$$V_t(A_t) = \sum_{k=1}^{n} e^{-r(T_k - t)} \mathbb{E}^{\mathbb{Q}} \left[\varphi_k^F(A_{T_k}) | A_t \right]$$

$$= \sum_{k=1}^{n} e^{-r(T_k - t)} \mathbb{E}^{\mathbb{Q}} \left[\left. \frac{\sigma_k^F}{\sqrt{T_k}} A_{T_k} + \mu_k^F \right| A_t \right].$$

We now utilize equation (12) so that

$$V_t(A_t) = \sum_{k=1}^{n} e^{-r(T_k - t)} \mathbb{E}^{\mathbb{Q}} \left[\frac{\sigma_k^F}{\sqrt{T_k}} \left(A_t \right. \right.$$

$$\left. \left. + \widehat{\kappa}(T_k - t) + \sqrt{T_k - t} Z \right) + \mu_k^F \right]$$

$$= \sum_{k=1}^{n} e^{-r(T_k - t)} \left(\frac{\sigma_k^F}{\sqrt{T_k}} \left(A_t + \widehat{\kappa}(T_k - t) \right) + \mu_k^F \right).$$

For the case where $t = 0$, $A_0 = 0$, therefore

$$V_0 = \sum_{k=1}^{n} e^{-rT_k} \left(\widehat{\kappa} \sigma_k^F \sqrt{T_k} + \mu_k^F \right)$$

and the proof is complete. □

Equations (10) and (11) provide a very simple alternative to valuing cash-flows using standard DCF approaches. The inherent risk of the project is captured by the cash-flow uncertainty estimates, rather than relying on an exogenously determined hurdle rate. We emphasize, that, while at first glance it may appear to be difficult to estimate ρ_{FI}, historical data of the market sector indicator to a traded asset can be utilized. Furthermore, if an assumption of ρ_{FI} is made, this is only a single assumption compared to using the β in the CAPM formulation. If one wanted to be consistent with the DCF CAPM based valuation, the ρ from the β could be backed out. A major advantage of our proposed approach is that two projects with the same expected cash-flow estimates will necessarily be valued differently if their estimated distributions are different. In our experience while working with managers, this was a very favorable aspect of our model.

C. Real Option Valuation of Risky Cash-Flows

In the real options context, similar to the scenarios proposed by many others (see for example Copeland and Antikarov (2001), Datar and Mathews (2004), Collan, Fullér, and Mezei (2009)), we assume that at some time T_0, managers can invest an amount K in a project and receive the uncertain managerial estimated cash flows at times T_k. The value of the real option at time $t < T_0$ can be computed under the risk-neutral measure $\mathbb{Q}$ as follows

$$RO_t(A) = e^{-r(T_0 - t)} \mathbb{E}^{\mathbb{Q}} \left[\left(V_{T_0}(A_{T_0}) - K \right)_+ \middle| A_t = A \right], \tag{13}$$

leading to the following Proposition.

Proposition 3. Real Option Value of Risky Cash-Flow Estimates. *For a given set of cash-flow estimates, normally distributed with mean μ_k^F and standard deviation σ_k^F, given at times T_k, where $k = 1, 2, ..., n$, the value of the option at time $t < T_0$ to invest the amount K at time $T_0 < T_k$ to receive these cash flows is given by*

$$RO_t(A_t) = e^{-r(T_0 - t)} \left[\left(\xi_1(A_t) - K \right) \Phi \left(\frac{\xi_1(A_t) - K}{\xi_2} \right) \right.$$

$$\left. + \xi_2 \phi \left(\frac{\xi_1(A_t) - K}{\xi_2} \right) \right] \tag{14}$$

where $\Phi(\bullet)$ and $\phi(\bullet)$ are the standard normal distribution and density functions, respectively, and

$$\xi_1(A_t) = \sum_{k=1}^{n} e^{-r(T_k - T_0)} \left(\frac{\sigma_k^F}{\sqrt{T_k}} \left(A_t + \widehat{\kappa}(T_k - t) \right) + \mu_k^F \right), \tag{15}$$

$$\xi_2 = \sqrt{T_0 - t} \sum_{k=1}^{n} e^{-r(T_k - T_0)} \frac{\sigma_k^F}{\sqrt{T_k}}. \tag{16}$$

At $t = 0$, ξ_1 reduces to

$$\xi_1 = \sum_{k=1}^{n} e^{-r(T_k - T_0)} \left(\widehat{\kappa} \sigma_k^F \sqrt{T_k} + \mu_k^F \right). \tag{17}$$

Proof. Using equation (10) with $t = T_0$ and substituting $V_{T_0}(A_{T_0})$ in equation (13) leads to

$$RO_t(A) = e^{-r(T_0 - t)} \mathbb{E}^{\mathbb{Q}} \left[\left(\sum_{k=1}^{n} e^{-r(T_k - T_0)} \right. \right.$$

$$\left. \left. \left(\frac{\sigma_k^F}{\sqrt{T_k}} \left(A_t + \widehat{\kappa}(T_k - t) + \sqrt{T_0 - t} Z \right) + \mu_k^F \right) - K \right)_+ \right]. \tag{18}$$

Defining

$$\xi_1(A_t) \equiv \sum_{k=1}^{n} e^{-r(T_k - T_0)} \left(\frac{\sigma_k^F}{\sqrt{T_k}} \left(A_t + \widehat{\kappa}(T_k - t) \right) + \mu_k^F \right),$$

$$\xi_2 \equiv \sqrt{T_0 - t} \sum_{k=1}^{n} e^{-r(T_k - T_0)} \frac{\sigma_k^F}{\sqrt{T_k}}$$

we have

$$RO_t(A) = e^{-r(T_0 - t)} \mathbb{E}^{\mathbb{Q}} \left[\left(\xi_1(A_t) - K + \xi_2 Z \right)_+ \right],$$

which simplifies to

$$RO_t(A_t) = e^{-r(T_0 - t)} \left[\left(\xi_1(A_t) - K \right) \Phi \left(\frac{\xi_1(A_t) - K}{\xi_2} \right) \right.$$

$$\left. + \xi_2 \phi \left(\frac{\xi_1(A_t) - K}{\xi_2} \right) \right]$$

completing the proof. □

The result of Proposition 3 is significant for it provides a simple analytical real options valuation method that is easy to apply and is consistent with financial theory. Arguably, it meets the requirements of reducing solution complexity, minimizing parameter estimation and being intuitive (see Block (2007)). We believe that this result can easily be embraced by practitioners.

IV. PRACTICAL IMPLEMENTATION

The theoretical foundation was established in Section III; here we provide a practical implementation of the methodology. We assume that a company is interested in investing in an early stage R&D project. The project will require a substantial investment of $50 million 2 years from now, at which point sales will begin and cash-flows will be realized in years 3 to 10. The market parameters are assumed to be as follows:

- Risk-free rate: $r = 3\%$
- Expected market growth: $\mu = 9\%$
- Market volatility: $\sigma = 10\%$.

Managers have estimated the cash-flows to be as depicted in Table I and the correlation of the cash-flows to the traded index are estimated to be 0.5. Furthermore, the managers have supplied standard deviation estimates for each of sales, cost of goods sold (COGS) and capital expenditures (CAPEX) as a percentage of each, as presented in Table II. As can be seen uncertainty increases farther out in the future. Furthermore, COGS are estimated to be 60% correlated to sales and CAPEX 50%. If we assume that the correlation between COGS and CAPEX is driven solely through their correlations to sales, then the standard deviation of the cash-flows can be calculated as follows,

$$\sigma_{CF} = \left(\sigma_S^2 + \sigma_C^2 + \sigma_{EX}^2 - 2\rho_{S,C}\sigma_S\sigma_C \right.$$
$$\left. -2\rho_{S,EX}\sigma_S\sigma_{EX} + 2\rho_{C,EX}\sigma_C\sigma_{EX}\right)^{\frac{1}{2}} \quad (19)$$

where σ denotes the standard deviation, ρ denotes the correlation and the subscripts "S", "C" and "EX" represent sales, COGS and CAPEX, respectively. The standard deviation values are provided in Table III.

TABLE I
MANAGERIAL SUPPLIED CASH-FLOW (MILLIONS $).

	3	4	5	6	7	8	9	10
Sales	10.00	30.00	50.00	100.00	100.00	80.00	50.00	30.00
COGS	6.00	18.00	30.00	60.00	60.00	48.00	30.00	18.00
GM	4.00	12.00	20.00	40.00	40.00	32.00	20.00	12.00
SG&A	0.50	1.50	2.50	5.00	5.00	4.00	2.50	1.50
EBITDA	3.50	10.50	17.50	35.00	35.00	28.00	17.50	10.50
CAPEX	1.00	3.00	5.00	10.00	10.00	8.00	5.00	3.00
Cash-Flow	2.50	7.50	12.50	25.00	25.00	20.00	12.50	7.50

TABLE II
MANAGERIAL SUPPLIED STANDARD DEVIATION PERCENT ESTIMATES.

	3	4	5	6	7	8	9	10
σ_S (Sales) %	10%	11%	12%	13%	15%	16%	18%	19%
σ_C (COGS) %	10%	11%	12%	13%	15%	16%	18%	19%
σ_{EX} (CAPEX) %	5%	6%	6%	7%	7%	8%	9%	10%

Using equation (11), the present value of the cash-flows is calculated to be $58.8 million and using equation (17) in equation (14) with $t = 0$ gives a value of $16.1 million for the real option.

TABLE III
ESTIMATED STANDARD DEVIATIONS (MILLIONS $).

	3	4	5	6	7	8	9	10
σ_S (Sales) 1.00	3.30	6.05	13.31	14.64	12.88	8.86	5.85	
σ_C (COGS)	0.60	1.98	3.63	7.99	8.78	7.73	5.31	3.51
σ_{EX} (CAPEX)	0.05	0.17	0.30	0.67	0.73	0.64	0.44	0.29
σ_{CF} (Cash-Flow)	0.78	2.58	4.73	10.40	11.44	10.07	6.92	4.57

V. CONCLUSIONS

In this work we developed a method to value cash-flows and the real option value of the cash-flows that requires only the solution of a simple analytical formula. The model is consistent with financial theory and properly accounts for both idiosyncratic and systematic risk. Through the introduction of a market sector indicator, which is used to drive the cash-flows, the methodology provides a natural mechanism to link the cash-flows from one period to the next. Furthermore, because the market sector indicator is assumed to be partially correlated to a traded index, we are able to develop a valuation under the risk-neutral minimum martingale measure. The method requires minimal parameter estimation, as only the correlation between the market sector indicator and the traded index may not be easily obtained, however, we believe, it can often be estimated based on historical data. Managers especially like the intuitive aspect of the model where the value of the cash-flows is linked to their estimated uncertainty. As was discussed in Jaimungal and Lawryshyn (2010), depending on the value of the correlation of the market sector indicator to the market index, and the strike cost, K, the method will lead to a decreasing or increasing real option value as cash-flow uncertainty is increased. This feature is realistic, however is not often properly accounted for in other real options methods, because many other real options approaches presented in the literature assume the value of the cash-flows can be perfectly hedged with a traded asset – i.e. they do not allow for idiosyncratic risk. Ultimately, we believe that our proposed method preserves the essential elements to make it consistent with financial theory, yet simple enough to implement so that it can be easily embraced by practitioners not versed in financial mathematics.

REFERENCES

Berk, J., D. P. and D. Stangeland (2010). *Corporate Finance (Canadian Edition)*. Pearson Canada Inc.

Black, F. and M. Scholes (1973). The pricing of options and corporate liabilities. *Journal of Political Economy 81*, 637–659.

Block, S. (2007). Are real options actually used in the real world? *Engineering Economist 52*(3), 255–267.

Borison, A. (2005). Real options analysis: Where are the emperor's clothes? *Journal of Applied Corporate Finance 17*(2), 17–31.

Brando, L. E., J. S. Dyer, and W. J. Hahn (2012). Volatility estimation for stochastic project value models. *European Journal of Operational Research 220*(3), 642–648.

Brennan, M. J. and S. Schwartz (1985). Evaluating natural

resource investments. *Journal of Business 58*(2), 135–157.

Cerny, A. (1999). Minimal martingale measure, capm, and representative agent pricing in incomplete markets. Technical report, Cass Business School. Available at SSRN: http://ssrn.com/abstract=851188.

Collan, M., R. Fullér, and J. Mezei (2009). A fuzzy payoff method for real option valuation. *Journal of Applied Mathematics and Decision Sciences*, 1–14.

Copeland, T. and V. Antikarov (2001). *Real Options: A Practitioner's Guide*. W. W. Norton and Company.

Copeland, T. and P. Tufano (2004, March). A real-world way to manage real options. *Harvard Business Review 82*(3), 90–99.

Datar, V. and S. Mathews (2004). European real options: An intuitive algorithm for the black- scholes formula. *Journal of Applied Finance 14*(1), 45–51.

Dixit, A. and R. Pindyck (1994). *Investment under Uncertainty*. Princeton University Press.

Föllmer, H. and M. Schweizer (1991). *Applied Stochastic Analysis : Stochastics Monographs*, Chapter Hedging of contingent claims under incomplete information, pp. 389–414. Gordon and Breach, London.

Jaimungal, S. and Y. A. Lawryshyn (2010). Incorporating managerial information into real option valuation. *Avaiable at SSRN: http://ssrn.com/abstract=1730355*.

Jaimungal, S. and Y. A. Lawryshyn (2011, June). Incorporating managerial information into valuation of early stage investments. In *Proceedings of the Real Options Conference*.

Myers, S. (1977). Determinants of corporate borrowing. *Journal of Financial Economics 5*, 147–175.

Pendharkar, P. (2010, March). Valuing interdependent multi-stage it investments: a real options approach. *European Journal of Operational Research 201*(3), 847–859.

Ryan, P. and G. Ryan (2002). Capital budget practices of the fortunre 1000: How things have changed? *Journal of Business and Management 8*(4), 355–364.

Smith, J. and R. Nau (1995, May). Valuing risky projects: Option pricing theory and decision analysis. *Management Science 41*(5), 795–816.

Trigeorgis, L. (1996). *Real Options: Managerial Flexibility and Strategy in Resource Allocation*. Cambridge, MA: The MIT Press.

Mapping Real Option Valuation Model Types to Different Types of Uncertainty

Mikael Collan

Lappeenranta University of Technology, School of Business
Lappeenranta, Finland
mikael.collan@lut.fi

Tero Haahtela

Aalto University, BIT Research Center
Helsinki, Finland
tero.haahtela@aalto.fi

Abstract—**This paper discusses different types of uncertainty and the fit of different real option valuation models to situations under the different types of uncertainty. A suitability-based mapping of real option valuation models for each type of uncertainty is made and discussed. Implications to decision-making are direct: wrong model selection may lead to unreliable results and poor decisions.**

Keywords—real option valuation; uncertainty; financial modelling, decision-making

I. INTRODUCTION

The focus of this short paper is to propose a mapping of (six different) real option valuation (ROV) model types to (three) types of uncertainty, based on a short analysis of the models and the types of uncertainty, and how well the models are able to take into consideration the type of information available under each type of uncertainty. This is important, because an unsuitable model-uncertainty combination will likely cause the obtained valuation results to be unreliable and may cause poor decisions. It is also rather clear from previous literature on real option valuation [1] that "one model does not fit all situations". Interestingly, there are very few, if any attempts to map real option valuation models' by usability to different types of uncertainty in the previous literature on real options.

Real options are real world choices and possibilities to make changes in projects and investments, before or during their economic life. These choices and possibilities are often referred to as managerial flexibility and they add value to projects: ceteris paribus, a project with real options is worth more than the same project without real options. How valuable real options are is an important question, because it enables cost-benefit analysis, when real options can be acquired and enhances decision-making, when selection between projects with real options is made.

The term "real option" is based on the observation that a real option has a strong similarity to a financial option; both offer their holder a choice, but not an obligation to act with regards to an underlying asset at a given future time, or during a given future period. ROV was originally based on the notion that as the analogy between real and financial options is strong, then financial option valuation models could be used

also for real option valuation. Six types of ROV models are presented in section II of this paper.

Uncertainty is something that is always present in future-oriented analysis, such as real option valuation. Uncertainty however is not "homogenous" in the sense that there are different types of uncertainty. In fact, it is commonplace that the more distant the future that must be evaluated is, the more difficult the analysis becomes. This increased difficulty is often caused by the increased complexity (increased number of the combinations of things, on different levels that may happen) connected to the future and the decreasing amount of information and knowledge that we have about the future. Three types of uncertainty are presented in section III of this paper.

Section IV of the paper maps the usability of the different types of ROV models to the different types of uncertainty, while section V is used for discussion and to draw some conclusions and closes the paper.

II. TYPES OF REAL OPTION VALUATION MODELS

It may be possible to classify ROV models in many ways (see e.g. [2, 3] for examples), here we have adopted the use of six classes of ROV models, based on the mathematical construct of the methods. Furthermore, we have selected a representative model for each model type to act as a background for closer inspection. The model types, the representative models selected, and a short introduction of each model type are given in the following:

i) Differential equation models, represented by the Black & Scholes option pricing formula from 1973 [4]. These methods express the option price and its dynamics using partial differential equations to relate the continuously changing value of the option to observable changes in market securities. The boundary conditions specify the option value at known points and its value at the extremes. Most of the continuous-time models assume the value of the underlying asset to follow a known stochastic process – most often geometric Brownian motion - and that the returns are normally distributed. Input into the models are numerous variables' values. Probability distribution of the underlying is considered to be exactly and objectively known based on available historical market data and complete markets with the opportunity for replication and hedging. Most applications in the field are related to natural resources investments - oil, natural gas, mining industry [5], where market data may be available for estimating the underlying asset process parameters.

ii) Discrete lattice models, represented by the Binomial Option Pricing model by Cox, Ross & Rubinstein from 1979 [6]. Lattice valuation models are based on a simple discrete representation of the evolution of the underlying asset value. Theoretical valuation assumptions are very similar to those for differential equation models, but the technical requirements are less restrictive. Lattices are much more easily explained to and accepted by management because the methodology is much simpler to understand [7]. Lattices are valuable especially with sequential and parallel compound options, which is often the case in real applications [8, 9]. They allow valuation of American options with early exercise possibility and they are suitable for valuing barrier options and derivatives dependent of several underlying assets [10, 11]. The approach can be a applied to several stochastic processes, including a mean-reverting process for valuing an enhanced oil recovery technology [12].

iii) Decision tree based models for valuing sequential investments and other investments with mutually exclusive alternatives. Decision tree analysis involves forecasting future outcomes and assigning probabilities to those events. Using DTA for real option analysis is very similar to the common discounted cash-flow based trees. However, instead of using actual probabilities, cash-flows are weighted using 'risk-neutral' probabilities derived from option pricing. This allows using risk-free interest rate to discount the cash-flows to present value [13]. The advantage of the DTA based approach is that it provides a comprehensive overview for the alternative scenarios of a decision. The method is also more intuitive to understand and easier to model than binomial lattices. DTA approach does not require stochastic process, and it can accommodate even a great deal of uncertainty about the future. In case of several overlapping and parallel options the DTA method is not as flexible as lattice based approaches. Especially common R&D stage-gate models, pharmaceutical valuation models [13], and other sequential investments are suitable for DTA real option approach (see e.g. [14] example of oil investment). Smith & Nau (1995) [15] approach is suitable for such cases where market and technical uncertainty are kept separate. A decision tree without precise cash-flow estimates and estimated probabilities can be used for explorative mapping of the possible outcomes.

iv) Marketed asset disclaimer (MAD) models; an approach originally by Copeland & Antikarov [9], where the project cash-flow itself is the underlying asset. Copeland and Antikarov suggest that the present value of the project cash-flows, without options (i.e., the traditional NPV less investment costs) is the best unbiased estimator of the market value of the project, were it a traded asset. According to this Marketed Asset Disclaimer, or MAD assumption, the project itself is the underlying asset of the replicating portfolio. Another assumption, based on Samuelsson's proof (1965) [16], is that the variations in the value of the project follow a random walk, thus justifying use of geometric Brownian motion as the stochastic process for the underlying asset. Along with the MAD assumption, Copeland & Antikarov also suggest a whole four-step valuation process framework for valuing investments with real options. Most distinctive property of this approach is that instead of observable market data, the approach relies also on managerial assumptions related to cash-flows. Valuation data is more subjective, and uncertainty is assumed to be both parametric and structural (yet they are not distinguished). Also market uncertainty and technical uncertainty are often treated similarly and combined together. However, the original approach has been extended by several authors, and the approach been applied for valuing R&D and pharmaceuticals [17].

v) Simulated cash-flow pay-off based models for option valuation, represented by the Datar – Mathews model for real option valuation from 2004 [18]. This group consists of approaches that use cash-flow simulation to determine the terminal value distribution and especially its pay-off distribution (or several distributions for different time periods). Then in the most simple case [19], the value of the option is defined as *Risk Adjusted Success Probability * (Benefits – Costs)* discounted to present value. Costs and revenues may have different discounting rates. The method is simple and intuitive, and does not have any assumption or requirement for a stochastic process, and the method is technically very robust. Both subjective and objective data can be used. Technical and market uncertainty may be combined, or kept separate. The method can be applied to nearly any kind of uncertain investment valuation as long as the cash-flow calculation parameter estimation is possible, at least on a somewhat reliable level. The approach has been applied, e.g. by the Boeing corporation [19] for R&D evaluation.

vi) Fuzzy pay-off based models, represented by the fuzzy pay-off method (FPOM) for real option valuation by Collan, Fullér, and Mezei from 2009 [20, 21]. These methods are based on the idea that through creation of multiple scenarios (minimum and maximum possible, and in between these two extremes) for future cash-flows, it is possible to simplify reality and to create a pay-off distribution. The pay-off distribution is created by mapping the NPV of the generated scenarios (usually three or four) to form either a triangular or a trapezoidal distribution of the NPV. The pay-off distribution is then treated as a fuzzy number and real option value is computed directly from the distribution. The models use the ROV logic, but calculation is done by using fuzzy arithmetic. The FPOM method is intuitive and simplistic, directly usable with spread-sheet software. Results can be easily presented graphically. The method can use both subjective and objective data in the creation of the cash-flow scenarios and is not dependent on any pre-selected process. The approach has been applied, e.g., for the valuation of future oriented assets, such as patents [22, 23] and R&D projects [24, 25] .

III.   TYPES OF UNCERTAINTY

The typology of uncertainty used here is adopted from Kyläheiko et al. [26] who divide uncertainty into three major classes:

i) Risk. In the footsteps of Knight [27] we will speak about *risk*, when the probability future events are objectively known, or at the least knowable. It is assumed that the probabilities of events are independent from the choice and actions.

ii) Parametric uncertainty. When a decision maker has certain knowledge of the structure of the problem (future), but is uncertain with regards to the parameters (probability) of the problem, the situation is that of *parametric uncertainty*. In other words the agent has an exhaustive list of the actions he may engage in, of all the possible states of the world, and of the consequences generated by his actions and the states of the world, but he has only subjective degree of beliefs with regards the probabilities of the occurrence of each state of the world [26].

iii) Structural uncertainty. *Structural uncertainty* is based on imperfect knowledge about the structure the future can take. The implications of structural uncertainty are at least the following: the set of possible relevant consequences of actions cannot be known a priori, utility functions may remain obscure, there may be unintended and unknown technological and organizational consequences of actions, and subjective probabilities of events may depend upon actions, which means that these endogenized events can no longer be like (pre-described) states of the world [28].

IV.   MAPPING TYPES OF UNCERTAINTY TO ROV MODELS

Here the six types of ROV models are mapped to the three types of uncertainty by considering information requirements of the model types and the structure (mathematical choices) of the model. The idea is to gain an understanding about the usability of the different ROV model types under the different types of uncertainty. Table I shows the mapping of the ROV model types to the different types of uncertainty with a short indication of the usability of the model for the types of uncertainty.

TABLE I.   MAPPING TYPES OF ROV MODELS TO TYPES OF UNCERTAINTY

Types of Real Option Models	Types of uncertainty		
	Risk	*Parametric uncertainty*	*Structural uncertainty*
Differential equation models	**Usable,** when the underlying stochastic process is identified and matched with a model. Needed input information is available and the model results are credible.	Usability compromised by the restrictive assumptions of the models and lack of available information.	Unusable. Information not available to determine input values reliably
Discrete lattice models	**Usable.** Needed information is available and model results are credible.	Usability compromised by the lack of available information.	Unusable. Information not available to determine input values reliably
Decision tree based models	**Usable** for mapping the relevant decisions and paths, precise valuation results possible.	**Usable** for mapping possible outcomes. Usability may be compromised for creating precise valuation results.	**Usable** for *exploring* different possibilities, not enough information for quantitative analysis.
Marketed asset disclaimer models	**Usable.** The relationship of the underlying with a marketed asset is fully known.	**Usable** in situations, where a relationship with a marketed asset or, e.g., an index can be reliably established.	Unusable, because the structure of the relationship between the underlying and any marketed assets is not known.
Simulated cash-flow models	**Usable,** the needed information is available to construct simulation models.	**Usable** generally, for as long as cash-flow calculation parameters can be reliably defined.	Unusable, because the structure of the problem is not understood – a simulation model cannot be reliably built and cash-flow calculation parameters cannot be defined.
Fuzzy pay-off based models	Technically usable, but design is not optimal for the situation.	**Usable** generally, based on managerial estimates for cash-flow scenarios, also when information is incomplete	**Usable** for creating *direction giving* quantitative results.

V.   DISCUSSION AND CONCLUSIONS

Mapping ROV model types to different types of uncertainty is important from the point of view of making coherent choices, with regards to model selection for analysis purposes. If a model that does not have the capacity to handle the type of situation for which it is selected, the results will most likely prove to be untrustworthy and may lead to erroneous (and costly) decisions. This paper has mapped six

ROV model types to three types of uncertainty in the attempt to clarify model usability.

This is to the authors' knowledge the first such mapping. It seems that there are only few ROV model types that are usable in analysis under structural uncertainty and that some model types seem to be usable only under risk type uncertainty.

REFERENCES

[1] A. Borison, "Real Options Analysis: Where Are the Emperor's Clothes?," *Journal of Applied Corporate Finance,* vol. 17, pp. 17-31, 2005.

[2] M. Collan, "Thoughts About Selected Models for Real Option Valuation," *Acta Universitatis Palackinae Olomucensis, Mathematica,* vol. 50, pp. 5-12, 2011.

[3] K. Barton and Y. Lawryshyn, "Integrating Real Options with Managerial Cash Flow Estimates," *The Engineering Economist,* vol. 56, pp. 254-273, 2011.

[4] F. Black and M. Scholes, "The Pricing of Options and Corporate Liabilities," *Journal of Political Economy,* vol. 81, pp. 637-659, 1973.

[5] G. Cortazar and J. Casassus, "Optimal timing of a mine expansion: Implementing a real options model," *The Quarterly Review of Economics and Finance,* vol. 38, pp. 755-769, 1998.

[6] J. Cox, *et al.*, "Option Pricing: A Simplified Approach," *Journal of Financial Economics,* vol. 7, pp. 229-263, 1979.

[7] J. Mun, *Real Options Analysis: Tools and Techniques for Valuing Strategic Investments and Decisions.* Hoboken, N.J., USA: John Wiley & Sons, 2002.

[8] L. Trigeorgis, *Real Options. Managerial Flexibility and Strategy in Resource Allocation.* Boston, Mass., U.S.: MIT Press, 1996.

[9] T. Copeland and V. Antikarov, *Real Options: a practitioner's guide.* New York, NY: Texere, 2001.

[10] P. Boyle, "A lattice framework for option pricing with two state variables," *Journal of Financial and Quantitative Analysis,* vol. 23, pp. 1-12, 1988.

[11] B. Kamrad and P. Ritchken, "Multinomial Approximating Models for Options with k State Variables," *Management Science,* vol. 37, pp. 1640-1653, 1991.

[12] W. Hahn and J. Dyer, "Discrete time modelling of mean-reverting stochastic processes for real option valuation," *European Journal of Operational Research,* vol. 184, pp. 534-548, 2008.

[13] A. Jägle, "Shareholder Value, Real Options, and Innovation in Technology-Intensive Companies," *R&D management,* vol. 29, pp. 271-287, 1999.

[14] L. Brandao, *et al.*, "Using Binomial Decision Trees to Solve Real-Option Valuation Problems," *Decision Analysis,* vol. 2, pp. 69-88, 2005.

[15] J. E. Smith and R. F. Nau, "Valuing risky projects: option pricing theory and decision analysis," *Management Science,* vol. 41, pp. 795-816, 1995.

[16] P. Samuelsson, "Proof That Properly Anticipated Prices Fluctuate Randomly," *Industrial Management Review,* vol. 6, pp. 44-49, 1965.

[17] O. Borrissiouk and J. Peli, "Real Option Approach for R&D Valuation: Case Study at Serono International S.A.," *The Financier* vol. 8, pp. 7-71, 2003.

[18] V. Datar and S. Mathews, "European real options: an intuitive algorithm for the Black Scholes formula," *Journal of Applied Finance,* vol. 14, pp. 7-13, 2004.

[19] S. Mathews, *et al.*, "A Practical Method for Valuing Real Options: The Boeing Approach," *Journal of Applied Corporate Finance,* vol. 19, pp. 95-104, 2007.

[20] M. Collan, *et al.*, "Fuzzy pay-off method for real option valuation," *Journal of Applied Mathematics and Decision Systems,* vol. 2009, 2009.

[21] M. Collan, *The Pay-Off Method: Re-Inventing Investment Analysis.* Charleston, NC, USA: CreateSpace Inc., 2012.

[22] M. Collan and M. Heikkilä, "Enhancing Patent Valuation with the Pay-Off Method," *Journal of Intellectual Property Rights,* vol. 16, pp. 377-384, 2011.

[23] M. Collan and K. Kyläheiko, "Forward-Looking Valuation of Strategic Patent Portfolios Under Structural Uncertainty," *Journal of Intellectual Property Rights,* vol. 18, pp. 230-241, 2013.

[24] M. Collan and P. Luukka, "Evaluating R&D Projects as Investments by Using an Overall Ranking from Four New Fuzzy Similarity Measure Based TOPSIS Variants," *IEEE Transactions on Fuzzy Systems,* vol. 21, pp. 1-11, 2013.

[25] F. Hassanzadeh, *et al.*, "A practical approach to R&D portfolio selection using fuzzy set theory," *IEEE Transactions on Fuzzy Systems* vol. 20, pp. 615-622, 2012.

[26] K. Kyläheiko, *et al.*, "Dynamic capability view in terms of real options," *International Journal of Production Economics,* vol. 80, pp. 65-83, 2002.

[27] F. Knight, *Risk, Uncertainty and Profit.* Boston: Hart, Schaffner & Marx, 1921.

[28] K. Kyläheiko, "Coping with technology: a study on economic methodology and strategic management of technology," D.Sc. (Econ. & Bus. Admin.), School of Business, Lappeenranta University of Technology, Lappeenranta, 1995.

A Credibilistic Approach to Risk Aversion and Prudence

Irina Georgescu
Department of Computer Science
and Economic Cybernetics
Academy of Economic Studies
Bucharest, Romania
Email: irina.georgescu@csie.ase.ro

Jani Kinnunen
IAMSR, Åbo Akademi
Turku, Finland
Email: jani.kinnunen@abo.fi

Abstract—Risk aversion and prudence are deeply studied topics in probabilistic risk theory. In this paper an approach of the two topics by credibility theory of B. Liu and Y. Liu is attempted. The risk situations are modeled by fuzzy variables and the indicators of risk aversion and prudence are defined in the context of credibilistic expected utility theory. Properties of these indicators are studied and approximate calculation formulas are established. Finally, one establishes a relationship between prudence and a credibilistic model of optimal saving.

I. INTRODUCTION

Probabilistic risk theory [1], [17] starts from the hypothesis that the risk is mathematically modeled by random variables. Important topics of risk such as risk aversion and prudence of an agent in front of a situation of uncertainty are analyzed in the framework of probabilistic expected utility theory [5], [15], [18]. Risk indicators are defined and expressed by the expected value and variance of the random variables which model the risk.

Probabilistic models are not always efficient in the treatment of uncertainty problems. In the literature there are attempts to tackle uncertainty phenomena by possibilistic models [4], [2], [8] or by credibilistic models [12], [13].

The purpose of this paper is to study risk aversion and prudence by credibility theory. We shortly describe the content of the paper.

In Section II we recall from [12], [13], [20] the meaning of credibilistic expected value and credibilistic expected utility, and some of their calculation formulas. As results we mention an approximation formula of credibilistic expected utility and a characterization theorem of convex (resp. concave) functions by a Jensen–type inequality with credibilistic indicators.

Section III defines the property of an agent of being credibilistically risk averse and characterizes this property by the concavity of the utility function. The notions of credibilistic utility premium and credibilistic risk premium as measures of credibilistic risk aversion are introduced. A formula for the approximate calculation of credibilistic risk premium and a Pratt–type formula [17] in credibilistic context are proved.

In Section IV the property of an agent of being prudent in front of a credibilistically modeled risk situation is introduced. One proves that the necessary and sufficient condition

for the credibilistic prudence is the non–negativity of the third derivative of the utility function. Then the credibilistic precautionary premium is defined, notion analogous to that studied by Kimball [10] in case of probabilistic prudence. For this indicator which measures the intensity of the credibilistic prudence an approximation formula and a comparison theorem in terms of Kimball's degree of absolute prudence [10] are proved.

II. CREDIBILISTIC EXPECTED UTILITY

The impact of risk on optimal saving was studied first by Leland [11] and Sandmo [19] by introducing the notion of precautionary saving. They proved that if the third derivative of consumer's utility function is positive. The converse was established by Kimball [10], this way the connection between precautionary saving and prudence in the sense of Kimball being done. In this section we will study a credibilistic model of optimal saving. We will define the notion of credibilistic precautionary saving; the positivity of this indicator signifies exactly the "prudence" of the consumer. The main theorem of the section establishes the equivalence between the formal definition of the prudence of a consumer from Section IV and the positivity of credibilistic precautionary saving.

Let $\Omega \subseteq \mathbf{R}$ be a set of states and $\mathcal{P}(\Omega)$ the power set of Ω. The elements of $\mathcal{P}(\Omega)$ are called events.

By [12], [13], a credibility measure on Ω is a function $Cr : \mathcal{P}(\Omega) \to [0,1]$ with the following properties:

(a) $Cr(\Omega) = 1$;

(b) If $A, B \in \mathcal{P}(\Omega)$ then $A \subseteq B$ implies $Cr(A) \leq Cr(B)$;

(c) For any $A \in \mathcal{P}(\Omega)$, $Cr(A) + Cr(A^C) = 1$;

(d) For any family $(A_i)_{i \in I}$ of subsets of Ω with the property $\sup_{i \in I} Cr(A_i) < \frac{1}{2}$ the equality $Cr(\bigcup_{i \in I} A_i) = \sup_{i \in I} Cr(A_i)$ holds.

A credibility space is a triple $(\Omega, \mathcal{P}(\Omega), Cr)$, where Cr is a credibility measure on Ω. A fuzzy variable is an arbitrary function $\xi : \Omega \to \mathbf{R}$. The membership function $\mu : \mathbf{R} \to [0,1]$ associated with ξ is defined by:

$$\mu(x) = \min(2Cr(\xi = x), 1) \text{ for any } x \in \mathbf{R}.$$

Let $\xi : \Omega \to \mathbf{R}$ be a fuzzy variable whose membership is $\mu : \mathbf{R} \to [0,1]$.

According to [12], [13], the credibility of an event $A \in \mathcal{P}(\Omega)$ can be expressed by

(1) $Cr(A) = \frac{1}{2}[1 + \sup_{x \in A} \mu(x) - \sup_{x \notin A} \mu(x)]$

The credibility distribution $\Phi : \mathbf{R} \to [0,1]$ of a fuzzy variable ξ is defined by [13]

(2) $\Phi(x) = Cr(\xi \leq x)$ for any $x \in \mathbf{R}$.

A function $\phi : \mathbf{R} \to [0,1]$ is called the credibility density function of ξ if

(3) $\Phi(x) = \int_{-\infty}^{x} \phi(t)dt$ for any $x \in \mathbf{R}$, $\int_{-\infty}^{\infty} \phi(t)dt = 1$.

By [12], [13], the credibilistic expected value $Q(\xi)$ of a fuzzy variable ξ has the form:

(4) $Q(\xi) = \int_{0}^{\infty} Cr(\xi \geq t)dt - \int_{-\infty}^{0} Cr(\xi \leq t)dt$

provided that the two integrals are finite.

Let $u : \mathbf{R} \to \mathbf{R}$ be a utility function. The credibilistic expected value $Q(u(\xi))$ [13] is introduced by:

(5) $Q(u(\xi)) = \int_{0}^{\infty} Cr(u(\xi) \geq t)dt - \int_{-\infty}^{0} Cr(u(\xi) \leq t)dt$

provided that the two integrals are finite.

Throughout the paper we assume that the credibility distribution Φ of ξ verifies

(6) $\lim_{x \to -\infty} \Phi(x) = 0$; $\lim_{x \to \infty} \Phi(x) = 1$.

In this case according to [13] we have

(7) $Q(\xi) = \int_{-\infty}^{\infty} xd\Phi(x)$

One can easily see that for any $a, b \in \mathbf{R}$, $Q(a\xi + b) = aQ(\xi) + b$.

If u is a monotonic utility function then by [20]

(8) $Q(u(\xi)) = \int_{-\infty}^{\infty} u(x)d\Phi(x)$

When the credibility density function ϕ of ξ exists the following holds

(9) $Q(u(\xi)) = \int_{-\infty}^{\infty} u(x)\phi(x)dx$

Lemma 2.1: Let g, h be two monotonic utility functions and $\alpha, \beta \in \mathbf{R}$. If $u = \alpha g + \beta h$ then $Q(u(\xi)) = \alpha Q(g(\xi)) + \beta Q(h(\xi))$.

The following result appears implicitly in the proof of Proposition 8.7 from [9], but it has an intrinsic importance.

Proposition 2.2: Assume that the utility function u has the class C^2. Then for any $x \in \mathbf{R}$

(10) $Q(u(x + \xi)) \approx u(x + Q(\xi)) + \frac{u''(x+Q(\xi))}{2} \int_{-\infty}^{\infty} (t - Q(\xi))^2 d\Phi(t)$

Proof: By the second–order Taylor approximation formula

$u(x + t) = u((x + Q(\xi)) + (t - Q(\xi)))$

$\approx u(x+Q(\xi))+u'(x+Q(\xi))(t-Q(\xi))+\frac{1}{2}u''(x+Q(\xi))(t-Q(\xi))^2$

Using formula (8) it follows:

$Q(u(x + \xi)) \approx u(x + Q(\xi)) \int_{-\infty}^{\infty} d\Phi(t) + u'(x + Q(\xi)) \int_{-\infty}^{\infty} (t - Q(\xi))d\Phi(t) + \frac{1}{2}u''(x + Q(\xi)) \int_{-\infty}^{\infty} (t - Q(\xi))^2 d\Phi(t)$

But $\int_{-\infty}^{\infty} d\Phi(t) = 1$ and $\int_{-\infty}^{\infty} (t - Q(\xi))d\Phi(t) = \int_{-\infty}^{\infty} td\Phi(t) - Q(\xi) = 0$, thus

$Q(u(x + \xi)) \approx u(x + Q(\xi)) + \frac{u''(x+Q(\xi))}{2} \int_{-\infty}^{\infty} (t - Q(\xi))^2 d\Phi(t)$ ∎

Proposition 2.3: If the utility function u is continuous and monotonic then the following are equivalent:

(i) u is convex;

(ii) $u(Q(\xi)) \leq Q(u(\xi))$ for any fuzzy variable ξ.

Proof: $(i) \Rightarrow (ii)$ If u is convex then

$u(Q(\xi)) = u(\int_{-\infty}^{\infty} xd\Phi(x)) \leq \int_{-\infty}^{\infty} u(x)d\Phi(x) = Q(u(\xi))$

$(ii) \Rightarrow (i)$ Let a, b be two real numbers, $a < b$. We consider a fuzzy variable ξ with the credibility density:

$$\phi(t) = \begin{cases} \frac{1}{b-a} & \text{for} \quad x \in [a, b] \\ 0 & \text{otherwise} \end{cases}$$

Then $Q(\xi) = \frac{a+b}{2}$ and $Q(u(\xi)) = \frac{u(a)+u(b)}{2}$, thus $u(\frac{a+b}{2}) = u(Q(\xi)) \leq Q(u(\xi)) = \frac{u(a)+u(b)}{2}$. Thus $u(\frac{a+b}{2}) \leq \frac{u(a)+u(b)}{2}$ for any $a, b \in \mathbf{R}$, from where, by [16], Theorem 1.1.4, p. 14, it follows that u is convex. ∎

Corollary 2.4: If the utility function u is continuous and monotonic then the following are equivalent:

(i) u is concave

(ii) $u(Q(\xi)) \geq Q(u(\xi))$ for any fuzzy variable ξ.

III. CREDIBILISTIC RISK AVERSION

We consider an agent having a continuous and monotonic utility function $u : \mathbf{R} \to \mathbf{R}$ and a fuzzy variable ξ representing a risk situation. Let Φ be the credibility distribution of ξ.

Definition 3.1: Let $x \in \mathbf{R}$ be the initial wealth of the agent. The *credibilistic utility premium* $w(x, \xi, u)$ associated with the triple (x, ξ, u) is defined by

(1) $w(x, \xi, u) = u(x) - Q(u(x + \xi))$

$w(x, \xi, u)$ is the credibilistic version of the probabilistic utility premium from [7].

We recall from [8] that the *credibilistic risk premium* $\pi(x, \xi, u)$ associated with the triple (x, ξ, u) is the solution of the equation:

(2) $Q(u(x + \xi)) = u(x + Q(\xi) - \pi(x, \xi, u))$

From (1) and (2) it follows

(3) $w(x, \xi, u) = u(x) - u(x + Q(\xi) - \pi(x, \xi, u))$

Proceedings of FORS40 - 2013 Lappeenranta, Finland - ISBN 978-952-265-435-9

Definition 3.2: We say that an agent with the utility function u is *credibilistically risk averse* if for any wealth level x and for any fuzzy variable ξ with $Q(\xi) = 0$, $Q(u(x + \xi)) \leq u(x)$.

Remark 3.3: If in Definition 3.2 we replace the inequality $Q(u(x + \xi)) \leq u(x)$ with the converse inequality $Q(u(x + \xi)) \geq u(x)$ (resp. with $Q(u(x + \xi)) = u(x)$) then the agent is credibilistically risk seeking (resp. credibilistically risk neutral).

Proposition 3.4: The following assertions are equivalent:

(i) The agent u is credibilistically risk averse;

(ii) u is concave.

Proof: According to Definition 3.2 and Corollary 2.4 the following assertions are equivalent:

- u is credibilistically risk averse;

- for any $x \in \mathbf{R}$ and the fuzzy variable ξ with $Q(\xi) = 0$, $Q(u(x + \xi)) \leq u(Q(x + \xi))$

- for any fuzzy variable ζ, $Q(u(\zeta)) \leq u(Q(\zeta))$

- u is concave. ■

Similarly one can prove the following result.

Proposition 3.5: The following assertions are equivalent:

(i) The agent u is credibilistically risk seeking (resp. credibilistically risk neutral).

(ii) u is convex (resp. linear).

Remark 3.6: According to Proposition 3.4 and [18], p. 8, an agent u is credibilistically risk averse iff u is concave iff the agent u is possibilistically risk averse.

Proposition 3.7: [9] Assume that the agent u has the class C^2 and $u' > 0$, $u'' < 0$. Then the credibilistic risk premium $\pi(x, \xi, u)$ can be approximated by

(4) $\pi(x, \xi, u) \approx -\frac{1}{2} \frac{u''(x + Q(\xi))}{u'(x + Q(\xi))} \int_{-\infty}^{\infty} (t - Q(\xi))^2 d\Phi(t)$

Proof: Applying the first order Taylor approximation we have:

(5) $u(x + Q(\xi) - \pi(x, \xi, u)) \approx u(x + Q(\xi)) - \pi(x, \xi, u)u'(x + Q(\xi))$

From (2), (5) and Proposition 2.2 one obtains

$u(x + Q(\xi)) - \pi(x, \xi, u)u'(x + Q(\xi)) \approx u(x + Q(\xi)) + \frac{u''(x + Q(\xi))}{2} \int_{-\infty}^{\infty} (t - Q(\xi))^2 d\Phi(t)$

from where (4) follows immediately. ■

According to [1], [17], the Arrow-Pratt index of the utility function u is defined by

(6) $r_u(x) = -\frac{u''(x)}{u'(x)}$, $x \in \mathbf{R}$.

Then (4) is written

(7) $\pi(x, \xi, u) \approx \frac{1}{2} r_u(x + Q(\xi)) \int_{-\infty}^{\infty} (t - Q(\xi))^2 d\Phi(t)$

Consider two agents with the utility functions u_1, u_2 of class C^2 and verifying $u_1' > 0, u_2' > 0, u_1'' < 0, u_2'' < 0$. We denote $r_1(x) = r_{u_1}(x)$ and $r_2(x) = r_{u_2}(x)$ for any $x \in \mathbf{R}$.

Applying (7) one obtains

(8) $\pi(x, \xi, u_1) \approx \frac{1}{2} r_1(x + Q(\xi)) \int_{-\infty}^{\infty} (t - Q(\xi))^2 d\Phi(t)$

(9) $\pi(x, \xi, u_2) \approx \frac{1}{2} r_2(x + Q(\xi)) \int_{-\infty}^{\infty} (t - Q(\xi))^2 d\Phi(t)$

Applying (8) and (9) we obtain the following credibilistic version of Pratt theorem [17]:

Theorem 3.8: Under the above conditions on u_1, u_2, the following conditions are equivalent:

(i) $\pi(x, \xi, u_1) \geq \pi(x, \xi, u_2)$ for any $x \in \mathbf{R}$ and for any fuzzy variable ξ;

(ii) $r_1(x) \geq r_2(x)$ for any $x \in \mathbf{R}$.

Remark 3.9: The credibilistic risk premium $\pi(x, \xi, u)$ is an indicator which measures the risk aversion of the agent u in front of credibilistic risk: the bigger this indicator is, the bigger the agent's risk aversion is. Theorem 3.8 shows that, as in the case of probabilistic risk, the comparison of aversions to credibilistic risk of two agents is expressed in terms of the Arrow–Pratt index.

IV. PRUDENCE IN A CREDIBILISTIC SETTING

The concept of "prudence" of an agent in front of a risk situation described by a random variable was defined in [10]. In this section we propose an approach to prudence by credibility theory. We will prove credibilistic versions of results of Kimball from [10].

Consider an agent with the utility function u of class C^2 and $u' > 0, u'' < 0$. A fuzzy variable ξ with the credibility distribution Φ will represent the risk situation.

Let x be the initial wealth, k a positive constant and ξ a zero–mean risk in credibilistic sense $(Q(\xi) = 0)$. We denote

(1) $S(x, k, \xi, u) = u(x - k) + Q(u(x + \xi)) - u(x) - Q(u(x - k + \xi))$

One notices that

(2) $S(x, k, \xi, u) = w(x - k, \xi, u) - w(x, \xi, u)$

Definition 4.1: The agent u is *credibilistically prudent* if $S(x, k, \xi, u) \geq 0$ for any triple (x, k, ξ) with the above significance.

The above definition of credibilistic prudence is analogous to the definition of probabilistic prudence in the form of [3], [6]. Instead of probabilistic expected utility of [3], [6], in Definition 4.1 the notion of credibilistic expected utility is used.

Remark 4.2: From (2) it follows that the agent u is credibilistically prudent iff the credibilistic utility premium $w(x, \xi, u)$ is decreasing in x.

Proposition 4.3: Assume that the utility function u has the class C^3 and $u' > 0, u'' < 0$. Then the agent u is credibilistically prudent iff $u''' \geq 0$.

Proof: We derive relation (1) of Definition 3.1:

(3) $w'(x, \xi, u) = u'(x) - Q(u'(x + \xi))$

Taking into account (3), Remark 4.2 and Corollary 2.4, the following assertions are equivalent:

- the agent u is credibilistically prudent;

- $w'(x, \xi, u) \leq 0$ for any x and ξ;

- $u'(x) \leq Q(u'(x + \xi))$ for any x and ξ;

- $u'(Q(x + \xi)) \leq Q(u'(x + \xi))$ for any x and ξ;

- u' is convex;

- $u''' \geq 0$.

■

Let u be a utility function of class C^3 with $u' > 0, u'' < 0, u''' > 0$, ξ a fuzzy variable and $x \in \mathbf{R}$.

The notion of credibilistic precautionary premium introduced by the following definition is the credibilistic analogue of (probabilistic) precautionary premium of Kimball [10].

Definition 4.4: The *credibilistic precautionary premium* $\lambda(x, \xi, u)$ associated with the triple (x, ξ, u) is the unique solution of the equation:

(4) $Q(u'(x + \xi)) = u'(x + Q(\xi) - \lambda(x, \xi, u))$

Remark 4.5: $\lambda(x, \xi, u)$ is the credibilistic risk premium $\pi(x, \xi, -u')$.

We recall from [10] the definition of the degree of absolute prudence:

(5) $P_u(x) = -\frac{u'''(x)}{u''(x)}$ for any $x \in \mathbf{R}$.

Taking into account Remark 4.5, from Proposition 3.7 and Theorem 3.8 it follows:

Proposition 4.6: $\lambda(x, \xi, u) \approx \frac{1}{2} P_u(x + Q(\xi)) \int_{-\infty}^{\infty} (t - Q(\xi))^2 d\Phi(t)$

Proposition 4.7: Let u_1, u_2 be two utility functions of class C^3 with $u_1' > 0, u_2' > 0, u_1'' < 0, u_2'' < 0, u_1''' > 0, u_2''' > 0$. The following assertions are equivalent:

(i) $\lambda(x, \xi, u_1) \geq \lambda(x, \xi, u_2)$ for any $x \in \mathbf{R}$ and for any fuzzy variable ξ;

(ii) $P_{u_1}(x) \geq P_{u_2}(x)$ for any $x \in \mathbf{R}$.

Proposition 4.8: Assume that u has the class C^3, $u' > 0, u'' < 0, u''' > 0$. The following assertions are equivalent:

(i) For any fuzzy variable ξ, the credibilistic risk premium $\pi(x, \xi, u)$ is decreasing in wealth: $x_1 \leq x_2$ implies $\pi(x_2, \xi, u) \leq \pi(x_1, \xi, u)$;

(ii) For all $x \in \mathbf{R}$, $P_u(x) \geq r_u(x)$ (prudence is larger than absolute risk aversion).

Proof: Let ξ be a fuzzy variable with the credibility distribution Φ. According to formula (2) from Section III:

$$u(x + Q(\xi) - \pi(x, \xi, u)) = Q(u(x + \xi))$$
$$= \int_{-\infty}^{\infty} u(x + t) d\Phi(t)$$

Deriving this equality and applying again formula (2) from Section III:

$$(1 - \pi'(x, \xi, u))u'(x + Q(\xi) - \pi(x, \xi, u)) =$$
$$= \int_{-\infty}^{\infty} u'(x + t) d\Phi(t)$$

$$= -\int_{-\infty}^{\infty} g(x + t) d\Phi(t)$$
$$= -Q(g(x + \xi))$$
$$= -g(x + Q(\xi) - \pi(x, \xi, g))$$

where $g = -u'$. From these equalities we obtain the following expression of $\pi'(x, \xi, u)$:

$$\pi'(x, \xi, u) = \frac{g(x + Q(\xi) - \pi(x, \xi, g)) - g(x + Q(\xi) - \pi(x, \xi, u))}{u'(x + Q(\xi) - \pi(x, \xi, u))}$$

But $u' > 0$ and g is strictly increasing, thus the following assertions are equivalent:

- $\pi(x, \xi, u)$ is decreasing in x;

- For all $x \in \mathbf{R}$, $\pi'(x, \xi, u) \leq 0$;

- For all $x \in \mathbf{R}$, $g(x + Q(\xi) - \pi(x, \xi, g)) \leq g(x + Q(\xi) - \pi(x, \xi, u))$;

- For all $x \in \mathbf{R}$, $\pi(x, \xi, g) \geq \pi(x, \xi, u)$.

Taking into account the equivalence of the previous assertions, Proposition 4.8 and that $P_u = r_g$, the following assertions are also equivalent:

- For any fuzzy variable ξ, $\pi(x, \xi, u)$ is decreasing in x;

- For any fuzzy variable ξ and $x \in \mathbf{R}$, $\pi(x, \xi, g) \geq \pi(x, \xi, u)$;

- For any $x \in \mathbf{R}$, $r_g(x) \geq r_u(x)$;

- For any $x \in \mathbf{R}$, $P_u(x) \geq r_u(x)$.

■

The previous proposition establishes a relationship between credibilistic risk aversion and prudence.

V. CREDIBILISTIC PRECAUTIONARY SAVING AND PRUDENCE

In this section we present a credibilistic model of precautionary saving. It is analogous to the well known probabilistic model (see e.g. [5], p. 95) and to the possibilistic model of optimal saving from [14]. The novelty of our model is that the lifetime utility is defined in terms of credibilistic expected utility from Section II.

The notion of credibilistic precautionary saving is introduced as a measure of the impact of credibilistic risk on the level of optimal saving. The main result of the section connects the credibilistic precautionary saving with the notion of credibilistic prudence formally defined in Section IV.

As a starting point we will use the probabilistic two-period model of precautionary saving in the form of [5], p. 95. This model is characterized by the following elements:

- $u(y)$ and $v(y)$ are the utility functions of the consumer in period 0, 1, respectively

- in period 0 there is a sure income y_0 and in period 1 there is an uncertain income modeled by the random variable $\tilde{y}$

- s is the level of saving (for period 0)

- r is the rate of interest for saving

Assume that the utility functions u and v have the class C^3 and $u' > 0$, $v' > 0$, $u'' < 0$, $v'' < 0$.

The expected lifetime utility of the model is

(1) $V(s) = u(y_0 - s) + M(v((1+r)s + \tilde{y}))$

This probabilistic model is transformed into a credibilistic model if we consider that the uncertain income from period 1 is modeled by a fuzzy variable ξ and the expected lifetime utility has the form:

(2) $W(s) = u(y_0 - s) + Q((v(1+r)s + \xi))$

The consumer's problem is to find that value of saving s for which the maximum of $W(s)$ is attained. Then we consider the following optimization problem:

(3) $\max_s W(s)$

One proves that W is a strictly concave function, therefore the optimal solution $s^* = s^*(\xi)$ of problem (3) is given by:

(4) $W'(s^*) = 0$

We derive (2):

(5) $W'(s) = -u'(y_0 - s) + (1+r)Q(v'((1+r)s + \xi))$

thus from (4) one obtains:

(6) $u'(y_0 - s^*) = (1+r)Q(v'((1+r)s^* + \xi))$

The uncertain model from above is associated with the following model without uncertainty: the uncertain income ξ is replaced by the sure income $Q(\xi)$. The lifetime utility of the second model is:

(7) $W_1(s) = u(y_0 - s) + v((1+r)s + Q(\xi))$

and its optimization problem gets the form:

(8) $\max_s W_1(s)$

W_1 is a strictly concave function and the optimal solution $s_1^* = s_1^*(Q(\xi))$ of (8) is given by:

(9) $W_1'(s_1^*) = 0$

In this case we have

(10) $W_1'(s) = -u'(y_0 - s) + (1+r)v'((1+r)s + Q(\xi))$

and the optimum condition (9) can be written:

(11) $u'(y_0 - s_1^*) = (1+r)v'((1+r)s_1^* + Q(\xi))$

We face now two models of optimal saving: in one of them risk is present , in another, it is not. For each of them we have a maximum level of optimal saving, s^* and s_1^*. The difference $s^* - s_1^*$ will be called *precautionary risk saving* and will measure the change of optimal saving induced by the credibilistic risk.

Theorem 5.1: Under the above conditions the following assertions are equivalent:

(i) $s^*(\xi) - s_1^*(Q(\xi)) \geq 0$ for any fuzzy variable ξ;

(ii) $v'''(x) \geq 0$ for any $x \in \mathbf{R}$;

(iii) The consumer with the utility function v is credibilistically prudent.

Proof: The equivalence $(ii) \Leftrightarrow (iii)$ is exactly Proposition 4.3. In order to prove $(i) \Leftrightarrow (ii)$ we notice that from (6) and (10) it follows:

$W_1'(s^*) = (1+r)[v'((1+r)s^* + Q(\xi)) - Q(v'((1+r)s^* + \xi))]$

W_1' is strictly decreasing, therefore according to the previous equality we have

$s^*(\xi) \geq s_1^*(Q(\xi))$ iff $W_1'(s^*(\xi)) \leq W_1'(s_1^*(Q(\xi))) = 0$

iff $v'((1+r)s^* + Q(\xi)) \leq Q(v'((1+r)s^* + \xi))$

Further, the equivalence $(i) \Leftrightarrow (ii)$ follows applying Proposition 2.3. ∎

Remark 5.2: The previous theorem shows that the consumer v is credibilistically prudent iff the precautionary saving $s^* - s_1^*$ is positive. By this we get to an intuitive meaning of the notion of prudence: a consumer is "prudent" if the optimal saving $s^*(\xi)$ in the presence of risk ξ is bigger than the optimal saving $s^*(Q(\xi))$.

VI. CONCLUSION

In this paper credibility theory is taken as a basis of risk phenomena modeling. Then the risk is a fuzzy variable and the risk indicators are approximated by credibilistic expected utility [12], [13], [20]. We enumerate the main results of the paper:

• the characterization of convex and concave functions by Jensen–type credibilistic inequalities;

• the definition of the notions of credibilistic risk aversion and credibilistic prudence and their characterization;

• the study of the notions of credibilistic risk premium and credibilistic precautionary premium as measures of risk aversion and prudence, respectively (approximate calculations, comparison theorems).

• the development of a credibilistic model of optimal saving and its connection with the notion of prudence introduced in Section IV.

A future paper will attempt a credibilistic approach to these topics when several risk parameters exist.

ACKNOWLEDGMENTS

The authors would like to thank the anonymous referees for useful comments on the first version of the paper.

REFERENCES

[1] K. J. Arrow, Aspects of the theory of risk bearing, Helsinki: Yriö Jahnssonin Säätiö, 1965

[2] C. Carlsson, R. Fullér, Possibility for decision, Springer, 2011

[3] D. Crainich, L. Eeckhoudt, On the intensity of downside risk aversion, Journal of Risk and Uncertainty, 36(3), 2008, 267–276

[4] D. Dubois, H. Prade, Possibility theory, Plenum Press, New York, 1988

[5] L. Eeckhoudt, C. Gollier, H. Schlesinger, Economic and Financial Decisions under Risk, Princeton University Press, 2005

[6] L. Eeckhoudt, H. Schlesinger, Putting risk in its proper place, American Economic Review, 96(1), 2006, 280–289

[7] M. Friedman, L. Savage, The utility analysis of choices involving risk, Journal of Political Economy, 56(4), 1948, 279–304

[8]   I. Georgescu, Possibility theory and the risk, Springer, 2012

[9]   I. Georgescu, Computing the risk indicators in fuzzy systems, Journal of Information Technology Research, $\underline{5}$(4), October-December 2012, 63-84

[10] M. S. Kimball, Precautionary saving in the small and in the large, Econometrica, $\underline{58}$, 1990, 53–73

[11] H. Leland, Saving and uncertainty: the precautionary demand for saving, Quarterly J. Economics **82**, 1968, 465–473

[12] B. Liu, Y. K. Liu, Expected value of fuzzy variable and fuzzy expected value models, IEEE Transactions on Fuzzy Systems, $\underline{10}$, 2002, 445–450

[13] B. Liu, Uncertainty theory, Springer, Heidelberg, 2007

[14] A. M. Lucia Casademunt, I. Georgescu, Optimal saving and prudence in a possibilistic framework, Advances in Intelligent Systems and Computing, vol. 2017, 2013, 61–68

[15] A. Mas–Colell, M. D. Whinston, J. R. Green, Microeconomic theory, Oxford University Press, USA, 1995

[16] C. P. Niculescu, L. E. Perrson, Convex functions and their applications: A contemporary approach, Springer, 2005

[17] J. Pratt, Risk aversion in the small and in the large, Econometrica, $\underline{32}$, 1964, 122–130

[18] J. Quiggin, Generalized expected utility theory, Kluwer-Nijhoff, Amsterdam, 1993

[19] A. Sandmo, The effect of uncertainty on saving decision, Rev. Economic Studies **37**, 1970, 353–360

[20] F. Xue, W. S. Tang, R. Zhao, The expected value of a function of a fuzzy variable with a continuous membership function, Computers & Mathematics with Applications, $\underline{55}$, 2008, 1215-1224

Estimating the potential impact of efficiency improvement in reducing nutrient emissions

Natalia Kuosmanen
MTT Agrifood Research Finland
Economic Research
Helsinki, Finland

Timo Kuosmanen
School of Business
Aalto University
Helsinki, Finland

Abstract—*A new two-stage approach to modeling the nutrient emissions of crop farms is developed. In the first stage, we estimate the classic production function using farm-level data. We apply the stochastic nonparametric envelopment of data (StoNED) method that combines the axiomatic, nonparametric modeling of the production technology with a stochastic, probabilistic treatment of inefficiency and noise. Given the estimated input saving and output expansion potential for each farm in the sample, we estimate the impacts of the decreased fertilizer use and the increased crop yield on the nutrient balance of farms. This provides an aggregate level estimate on how much nutrient emissions could be abated through improving the operational efficiency of farms without any new technical innovations or down-scaling of the activity.*

Keywords—environmental performance; frontier estimation; productive efficiency analysis

I. INTRODUCTION

Productivity and efficiency analysis is frequently applied in the field of agricultural economics to estimate the total factor productivity growth at the sector and farm levels, to decompose the productivity changes to components of technological progress, operational efficiency and scale efficiency, to identify efficiency improvement targets and best-practice benchmarks, and to estimate shadow prices for undesirable outputs such as nutrient leaching to water systems (see e.g., [1], [2], [3]). All different applications of productive efficiency analysis are based on the estimation of the production frontiers in one way or another.

The two main approaches to estimating the best-practice frontiers are the nonparametric data envelopment analysis (DEA: [4], [5]) and the parametric stochastic frontier analysis (SFA: [6], [7]). The main appeal of DEA lies in its nonparametric treatment of the frontier, which does not assume a particular functional form but relies on the general regularity properties such as monotonicity, convexity, and homogeneity. However, DEA attributes all deviations from the frontier to inefficiency, and completely ignores any stochastic noise in the data. The key advantage of SFA is its stochastic treatment of residuals, decomposed into a non-negative inefficiency term and a disturbance term that accounts for measurement errors and other random noise. However, SFA builds on the parametric regression techniques, which require an ex ante specification of the functional form.

Bridging the gap between SFA and DEA was recognized as an important research question already in the early 1990s. The emerging literature on semi/nonparametric stochastic frontier estimation has thus far mainly departed from the SFA side, replacing the parametric frontier function by a nonparametric specification that can be estimated by kernel regression or local maximum likelihood (ML) techniques. Fan et al. [8] and Kneip and Simar [9] were among the first to apply kernel regression to frontier estimation in the cross-sectional and panel data contexts, respectively. Fan et al. [8] proposed a two-step method where the shape of the frontier is first estimated by kernel regression, and the conditional expected inefficiency is subsequently estimated based on the residuals, imposing the same distributional assumptions as in standard SFA. Kneip and Simar [9] similarly use kernel regression for estimating the frontier, but they make use of panel data to avoid the distributional assumptions.

Banker and Maindiratta [10] were the first to consider ML estimation of the stochastic frontier model subject to nonparametric shape constraints regarding the frontier. Their monotonicity and concavity constraints are analogous to DEA, but solving the resulting complex ML problem has proved extremely difficult, if not impossible in practical applications. No reported applications of the Banker and Maindiratta's constrained ML method are known in the literature.

In a series of recent papers, [11], [12], [13], [14] and [15] develop a new encompassing framework referred to as stochastic nonparametric envelopment of data (StoNED). Specifically, they introduce a stochastic SFA-style composite error term to a nonparametric, DEA-style piecewise linear frontier. Similar to DEA, the StoNED approach does not assume any particular functional form for the frontier; it identifies the function that best fits the data from the family of continuous, monotonic increasing, concave functions that can be non-differentiable. Kuosmanen and Kortelainen [12] noted that both SFA and DEA can be constrained special cases of the more general StoNED model.

An earlier study [16] presented the first application of the StoNED method to the farm-level data from agricultural production, focusing on the environmental performance of a sample of Finnish dairy farms. The present paper extends the scope of the previous study in two important respects. First, we shift attention on crop farms specialized in wheat cultivation. Second, we elaborate the modeling of nutrient leaching. Some problems with the conventional approach of modeling nutrient surplus as an input or undesirable output are discussed. Further, an alternative two-stage method that avoids these problems is developed.

The rest of the paper is organized as follows. In the next section we introduce the theoretical model of production involving undesirable outputs. We then present the StoNED method that can be applied for estimating the theoretical model. We then describe our empirical data and present the estimation results. A brief summary and discussion of future research topics concludes the paper.

II. StoNED Model

The vector of inputs (including labor, capital, energy, and intermediate inputs) is denoted by $\mathbf{x}$ and the output is denoted by y. The production technology is represented by the *production function* $y = f(\mathbf{x})$. We assume that function f belongs to the class of continuous, monotonic increasing and globally concave functions. In contrast to the SFA literature, no specific functional form for f is assumed a priori; our specification of the production function proceeds along the nonparametric lines of the DEA literature.

The observed output y_i of firm i may differ from $f(\mathbf{x}_i)$ due to inefficiency and noise. We follow the SFA literature and introduce a composite error term $\varepsilon_i = v_i - u_i$, which consists of the inefficiency term $u_i > 0$ and the idiosyncratic error term v_i, formally,

$$y_i = f(\mathbf{x}_i) + \varepsilon_i = f(\mathbf{x}_i) - u_i + v_i, \quad i = 1,...,n. \qquad (1)$$

Terms u_i and v_i ($i = 1,...,n$) are assumed to be statistically independent of each other as well as of inputs $\mathbf{x}_i$. Furthermore, we follow the standard SFA practice and assume $u_i \underset{i.i.d}{\sim} \left| N(0, \sigma_u^2) \right|$ and $v_i \underset{i.i.d}{\sim} N(0, \sigma_v^2)$.

In model (1), the deterministic part (i.e., production function f) is defined analogous to DEA, while the stochastic part (i.e., composite error term ε_i) is defined similar to SFA. As a result, model (1) encompasses the classic SFA and DEA models as its constrained special cases. Specifically, if f is restricted to some specific functional form (instead of the class F_2), model (1) boils down to the SFA model by [6]. On the other hand, if we impose the restriction $\sigma_v^2 = 0$ and relax the distributional assumption concerning the inefficiency term, we obtain the single-output DEA model with an additive output-inefficiency, first considered by Afriat [17]. In this sense, both SFA and DEA can be seen as constrained special cases of model (1).

It is easy to write a theoretical model like (1); the main challenge is its estimation. In this paper we propose a new strategy to estimating model (1), referred to as *stochastic nonparametric envelopment of data* (StoNED). Our objective is to estimate the deterministic part of the model in a fully nonparametric fashion imposing a minimal set of assumptions,

in the spirit of DEA. Similar to DEA, we estimate the shape of the frontier by exploiting the regularity properties from the microeconomic theory (i.e., continuity, monotonicity, and concavity of f), free of any distributional assumptions or assumptions about the functional form of f or its smoothness. However, in the cross-sectional setting it is impossible to distinguish between inefficiency and noise without imposing some distributional assumptions. Having estimated the shape of function f, we make use of the standard distributional assumptions adopted from the SFA literature to estimate the expected location of the frontier f, and the firm-specific conditional expected values for the inefficiency term. In summary, the StoNED method consists of two-steps:

Step 1: Estimate the shape of function f by Convex Nonparametric Least Squares (CNLS) regression

$$\min_{\alpha,\beta,\varepsilon} \sum_{i=1}^{n} \varepsilon_i^2$$
$$y_i = \alpha_i + \beta_i' \mathbf{x}_i + \varepsilon_i$$
$$\alpha_i + \beta_i' \mathbf{x}_i \le \alpha_h + \beta_h' \mathbf{x}_i \quad \forall h, i = 1,...,n$$
$$\beta_i \ge \mathbf{0} \quad \forall i = 1,...,n$$

Step 2: Using residuals ε from the CNLS regression, estimate the variance parameters σ_u^2, σ_v^2 by using the method of moments or pseudolikelihood techniques, and compute the conditional expected values of inefficiency. Subsequently, the expected value μ and the parameters of the inefficiency and noise distributions can be estimated based on the regression residuals by the method of moments or pseudolikelihood techniques (see [12] for details). $\alpha + \beta = \chi$. (1)　　(1)

III. Modeling of Nutrient Emissions

In the literature of productive efficiency analysis, the conventional approach to modeling nutrient emissions is to treat the nitrogen surplus as an input variable or as an undesirable output, e.g., [18]. However, the conventional approach is problematic both in the DEA and SFA-type modeling. Firstly, DEA requires that the inputs and outputs are ratio-scale variables. However, the nitrogen surplus is only an interval-scale variable: it is measured as a difference of the inflow and outflow of nitrogen. A related issue is that the nitrogen surplus may be less than or equal to zero. Secondly, from the perspective of the traditional production model assumed in the SFA literature, the nitrogen surplus is not a factor of production: it is possible to produce output even if the nitrogen surplus is negative or zero. Further, increase in the nitrogen surplus does not necessarily have an increasing effect on the output. Due to these problems, in this study we consider an alternative approach to modeling the nutrient problem.

In crop farms, high nitrogen surplus is generally due to the excessive use of fertilizers. Fertilizer is a ratio-scale input variable whereas the nitrogen surplus is not. Therefore, we resort to a two-stage approach where we first estimate the classic production function that includes the fertilizer use as one of the inputs. In the second stage, we estimate the potential impact of efficiency improvement on the nitrogen balance of the farm. Note that efficiency improvement can influence the nutrient surplus in two ways. Firstly, decreasing the input use, including the fertilizer input, will decrease the inflow of nitrogen. Secondly, increasing the output (crop yield) will increase the outflow of nutrient. Both have a favorable effect on the nutrient balance. In the application to be considered next, we examine both the input decrease and output expansion scenarios.

IV. DATA

We next apply the StoNED method described in the previous section to the empirical data of Finnish crop farms specialized in wheat. The data set is extracted from the Farm Accountancy Data Network (FADN) database. The following variables were considered: output variable is yield of wheat (100kg/ha), input variables are labor (hr/ha), farm capital (€/ha), total energy cost (€/ha) and fertilizers (kgN/ha), and the environmental variable of interest is the nitrogen surplus (kg/ha). An overview of the key characteristics of the data is presented in Table 1 in the form of the descriptive statistics: mean, standard deviation, minimum and maximum values.

TABLE I.   DESCRIPTIVE STATISTICS FOR THE SAMPLE OF WHEAT FARMS; YEAR 2004, SAMPLE SIZE N=141.

Variable	Mean	St. Dev.	Min	Max
Yield of wheat, 100kg/ha	36.6	11.3	7.2	71.6
Labor, hr/ha	29.0	17.4	5.8	117.8
Farm capital, €/ha	2853.5	1302.1	977.1	7918.9
Energy, €/ha	89.9	41.3	0.0	279.9
Fertilizer, kgN/ha	135.8	62.6	0.0	385.3
Nitrogen surplus kgN/ha	26.9	12.3	-0.1	77.1

V. MAIN RESULTS

We first applied the StoNED method to estimate efficiency in input use at the farm level. Recall that the StoNED method allows us to distinguish whether deviations from the frontier that are due to inefficiency or random noise. The estimated standard deviation of inefficiency was 11.6, whereas the estimated standard deviation of noise was 7.82. Both standard deviations were found to be statistically significant at the one percent significance level.

The expected inefficiency loss was estimated as 9.25 €/ha. This is approximately 20% of the average wheat revenue per hectare. Thus, expressed in relative terms, the average efficiency at the sample level was approximately 80%.

Next, we estimated the potential impact on the nitrogen surplus that would be achievable through improvement of technical efficiency. Specifically, suppose that all farms in the sample could catch up to the level of performance shown by the best farms in the sample. This would not require any new technical innovation, just that all farms could operate according to the current best practice. This is of course a very ambitious target as such, and unlikely to be met in practice. Nevertheless, it is interesting to examine what would be the impact of the maximum possible efficiency improvement on the nutrient balance. If the efficiency improvement does not suffice to bring

the nutrient surplus to a sustainable level, the further decrease must come either through technical progress or down-scaling production. In this respect, utilizing the efficiency improvement potential to the maximum extent might be the easiest and the most cost-efficient way to decrease nutrient leaching.

Efficiency improvement could be implemented by producing the current level of output by less input, or using the current input to produce a larger amount of output. In the case of crop farms, both decrease of fertilizer input and the greater yield of output have a favorable effect on the nutrient balance. Thus, both input and output orientations were considered.

In the output orientation, the impact comes through larger uptake of nitrogen in the harvested crop keeping the land use and fertilizers use at the present level. This would result only 1.9 kg/ha decrease in nitrogen surplus. In the input orientation, we keep the present output level constant and decrease the input use including land and fertilizer, which would result in 5.4 kg/ha decrease in nitrogen surplus. While any kind of efficiency improvement has a desirable effect, the direction of input saving or output expansion matters as well. A more favorable effect is achieved through input-oriented efficiency improvement.

In conclusions, there is some scope for decreasing nitrogen surplus through efficiency improvement, but it is rather limited. Our results show that the input orientation allows for larger improvement potential. Even though it is not at all clear what would be the sustainable level of nitrogen surplus, at least, it is possible to estimate how much emissions could potentially be decreased if all farms operated as efficiently as their most productive peers.

VI. CONCLUSIONS

We have explored a new approach to modeling and estimating the potential impact of efficiency improvement in reducing the nutrient emissions. The novel features of the study include the use of the StoNED method to estimating the frontier production function, and the subsequent estimation of the nutrient balance based on the decrease of fertilized input or increase of crop yield achievable through efficiency improvement. The attractive feature of the StoNED approach is that it enables us to combine the nonparametric, axiomatic approach to modeling the production technology with the stochastic modeling of inefficiency and noise. Further, drawing distinction between the productive input (fertilizer) and the associated environmental impact (nutrient surplus) avoids the theoretical and technical problems of the conventional modeling of nutrient surplus as an input or output variable.

Although we recognize that some environmental problems may depend on multiple inputs and/or outputs in a more complex way, we believe the two-stage approach considered in this paper has potential applications in other areas as well. For example, the green-house gas (GHG) emissions arising in the manufacturing sector are mainly due to one specific input: the energy use. Thus, a similar approach could be applied to estimating the potential reduction of GHG emissions through improving productive efficiency in the manufacturing industries.

REFERENCES

[1] T.J. Coelli, "Recent developments in frontier modelling and efficiency measurement," Australian Journal of Agricultural Economics, vol. 39(3), pp. 219-245, 1995.

[2] A. Oude Lansink, K. Pietola and A. Bäckman, "Effciency and productivity of conventional and organic farms in Finland 1994 – 1997," European Review of Agriculture Economics, vol. 29(1), pp. 51-65, 2002.

[3] M. Gorton and A. Davidova, "Farm productivity and efficiency in the CEE applicant countries: a synthesis of results," Agricultural Economics, vol 30, pp. 1-16, 2004.

[4] M.J. Farrell, "The measurement of productive efficiency," Journal of the Royal Statistical Society," Series A, vol. 120, pp. 253–281, 1957.

[5] A. Charnes, W.W. Cooper and E. Rhodes, "Measuring the inefficiency of decision making units," European Journal of Operational Research, vol. 2(6), pp. 429-444, 1978.

[6] D.J. Aigner, C.A.K. Lovell and P. Schmidt, "Formulation and estimation of stochastic frontier models," Journal of Econometrics, vol. 6, pp. 21-37, 1977.

[7] W. Meeusen and J. van den Broeck, "Efficiency estimation from cobb-douglas production function with composed error," International Economic Review, vol. 8, pp. 435-444, 1977.

[8] Y. Fan, Q. Li and A. Weersink, "Semiparametric estimation of stochastic production frontier models," Journal of Business and Economic Statistics, vol. 14(4), pp. 460-468, 1996.

[9] A. Kneip and L. Simar, "A general framework for frontier estimation with panel data," Journal of Productivity Analysis, vol. 7, pp. 187-212, 1996.

[10] R.D. Banker and A. Maindiratta, "Maximum likelihood estimation of monotone and concave production frontiers," Journal of Productivity Analysis, vol. 3, pp. 401-415, 1992.

[11] T. Kuosmanen, "Stochastic nonparametric envelopment of data: Combining virtues of SFA and DEA in a unified framework," MTT Discussion Paper 3/2006.

[12] T. Kuosmanen and M. Kortelainen, "Stochastic non-smooth envelopment of data: semi-parametric frontier estimation subject to shape constraints," Journal of Productivity Analysis, vol. 38(1), pp. 11-28.

[13] T. Kuosmanen, "Representation theorem for convex nonparametric least squares," Econometrics Journal, vol. 11, pp. 308-325, 2008.

[14] T. Kuosmanen and A.L. Johnson, "Data envelopment analysis as nonparametric least-squares regression," Operations Research, vol. 58(1), pp. 149-160, 2010.

[15] A.L. Johnson and T. Kuosmanen, "One-stage estimation of the effects of operational conditions and practices on productive performance: asymptotically normal and efficient, root-n consistent stonezd method," Journal of productivity analysis, vol. 36(2), pp. 219-230, 2011.

[16] T. Kuosmanen and N. Kuosmanen, "Role of benchmark technology in sustainable value analysis: an application to Finnish dairy farms," Agricultural and Food Science, vol. 18(3-4), pp. 302-316, 2009.

[17] S. Afriat, "Efficiency estimation of production functions," International Economic Review, vol. 13, pp. 568-598, 1972.

[18] S. Reinhard, C.A.K. Lovell and G. Thijssen, "Analysis of environmental efficiency variation," American Journal of Agricultural Economics, vol. 84(4), pp. 1054-1065, 2002.

A model for estimating and visualizing duration and time evolution of capability through decision points

Juha-Pekka Nikkarila
Defence Forces Technical Research Centre
Riihimäki
Finland

Juhani Hämäläinen
Defence Forces Technical Research Centre
Riihimäki
Finland

Abstract—Capabilities are needed for achieving profits in competitive situations. We shall here present a schematic model how duration of capability can be analyzed and visualized in discrete decision making process. The duration time of capability will be defined by the condition that capability vanishes and the consumption of capability is due to decisions provided at decision points. We introduce a function, which maps duration times to achieved profits and remaining capability values, to consider goodness of applied decisions. These quantities are useful in comparing different decision sequences with respect to corresponding capability duration and achieved profits. In order to improve given decision sequence, we consider time derivative of capability (value) and in particular its minimum values i.e. to identify corresponding time instants or intervals. The minimum values provide information of the fastest decreases of capability and they might be useful in identifying the reasons for that. The introduced procedure is applicable if the consumption of capabilities due to different decision is known or estimated.

I. INTRODUCTION

Capability is a widely used term for describing an ability to perform actions. An organization may use its capabilities for achieving its strategic goals. In real world, competing organizations may have conflicting interests. Two companies sharing the same markets is an example of conflicting interests. Both firms try to make a profit and e.g. losing customers would be considered as a drawback to the business. Furthermore, a loss of a customer to one company might be a gain of a customer to another company. Consequently, both companies use their capabilities and compete with each other in order to gain competitive advantage. In economics, capabilities are usually divided into financial, strategic, technological and organizational capabilities. [1], [2], [3]. Dynamic capability has been introduced and defined to help to understand time dependent challenges [4], [5], [6], [7].

Different nations may also have conflicting interests, but usually one achieves satisfactory results via diplomacy. Nations tend to design their defense organizations by estimating what type of threats the nation may encounter; creating and developing its military capabilities [8], [9] for handling the threats [10], [11]. A state may have to use also its military capability in order to achieve its strategic (and political) targets [12], [13]. One may consider humanitarian interventions as an example of using military capability for political reasons [14]. Usually there is an ally of nations or e.g. the UN or NATO conducting humanitarian interventions since one has to gain a strong political authority to the operation [15].

In this study, we discuss on modeling of capabilities' duration in situations where capabilities will vanish within finite time. The organization uses its capability for improving its competitive position compared to its competitors. The using strategy will be modeled through decisions, which determines the capabilities evolution. In other words, the capability is decided to use for gaining profit to its owner organization or for decreasing the conflicting organization's capabilities.

Modeling outcomes at competitive situations is traditionally provided by Game Theory. However, Game Theory provides profits or outcomes, but we focus on time evolution of capability during sequential decisions. In a sense Game Theory is intrisically present in our model since the opponent's choices affects on the decisions. Consequently, dynamic programming can be used in resolving the maximum duration of the capability. It is understood that the capability duration is not likely the only decision variable. Therefore, we introduce also other mathematical tools for improving the analysis.

II. RESOLVING AND PROLOGING MAXIMUM DURATION OF A CAPABILITY

We concentrate on capabilities' duration, which is aimed to be maximized and used effectively with respect to gained profits. We consider on capabilities decreasing as a function of time due to their usage and sudden losses. The usage is determined by owners' decisions at decisions points and the unexpected occasions affecting on capability cause sudden losses. The decisions guiding the use of capability are aimed to achieve best possible outcome in order to reach one's strategic goal. However, the using environment of capabilities is often shared with competing parties using their capabilities which affect on the profits or may be the reasons for the mentioned sudden losses. In the long term, the other party may benefit if the capability under interest is decreased and consequently, it can use its own capabilities in order to benefit itself. As a conclusive result, there can be sudden losses in the considered capability function.

Usually the owner of the capability encounters several decision-making situations where it is decided how and how much of the capability is used in the near future and how its usage is focused. At each decision-making situation there are several options of how the capability could be used. The usage of the capability influences also to the competitors and may decrease their capabilities as well. Since the competitor is affected by the usage of the capability it is less capable on using its own capabilities against the capability under interest.

As a result, a decision made at each decision-making situation leads to a unique time dependence of the capability.

A. Determining maximum duration of a capability

Let us define the capability function (over time) as $C(t)$ and relative to the initial capability (i.e. percentage value). At the decision points, there are several alternative choices for capability use in near future. The possible decisions determine the capability consumption, which can be modeled with different decreasing functions. Therefore, each decision point will provide new functions for capability consumption and the overall situation looks like a tree. By considering capability consumption through all included decision points, alternative durations of capability can be observed. We call these function graphs to capability paths. The different capability paths are described in the schematic Figure (1).

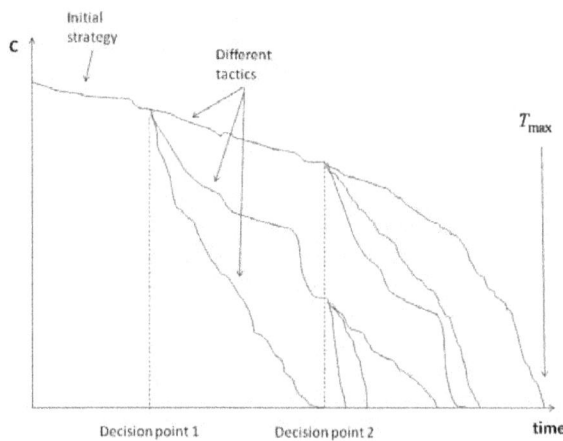

Fig. 1. Visualizing of the maximal duration of a capability. The alternative decisions at each decision point leads to a different capability time evolution, which make branches the capability paths at every decision point.

The maximum duration of a capability is dependent on the using decisions, environment and competing capabilities. The duration of the capability at scenario i can be visualized by capability paths and it can be given by the condition

$$t_{\mathrm{dur,i}} = \min\{t > 0 | C(t) = 0\}, \qquad (1)$$

where $C(t)$ is the capability time evolution obtained by i:th sequence of decisions. The maximum duration T_{max} provides maximum $t_{\mathrm{dur,i}}$ over the possible capability paths and it can be given by the following equation

$$T_{\mathrm{max}} = \max_i\{t_{\mathrm{dur,i}}\}. \qquad (2)$$

The visualization emphasizes on time dependence and duration of the capability. However, the analysis of capabilities is a complex process which should be conducted before the visualization can be done. The status of the capability is a function of competitors' or other parties' (op) capability $C_{\mathrm{op}}(t)$ and even more importantly, a function of their interaction $F(C(t), C_{\mathrm{op}}(t))$. In other words, the tactics or the strategy (decisions) on using the capability depends also on competitors' capabilities and interactions of capabilities.

Moreover, the interaction may cause unexpected and sudden changes (decreases) into the capability value. Finally, the time evolution as well as duration of the capability depends on several external variables.

Since the capability time evolution curve is not trivial, but rather complex, all the possible paths should be taken into account in order to resolve capability time dependency. Also, the interactions with other capabilities should be identified and analyzed. In other words, it is claimed that the only possibility to find out the capability time dependency is to systematically (experimentally) go through all the variations in the decision-making process. The Equation (2) gives a maximum duration of a capability and corresponding capability path.

The durations of capability are emphasized in the Figure (1) with the corresponding paths due to sets of decisions. The maximum duration and the corresponding set of decisions correspond to the goal of long duration of specific capability. However, the maximum duration does not necessarily correspond the maximum profits obtained by using the considered capability.

B. Comparing duration time of the capability to the obtained profits and the values of remaining capabilities

We have introduced a model of how to find the tactics or strategy (sequential decisions) maximizing the duration of a capability. However, the duration alone does not provide enough information on the goodness of decisions. In particular, the results the organization has gained by using its capability should be considered. In order to proceed, we introduce a function where the final states (of remaining capabilities or profits) are expressed at the moment when the considered capability vanishes i.e. at the duration time of the capability. This provides information of values of decisions made along different paths.

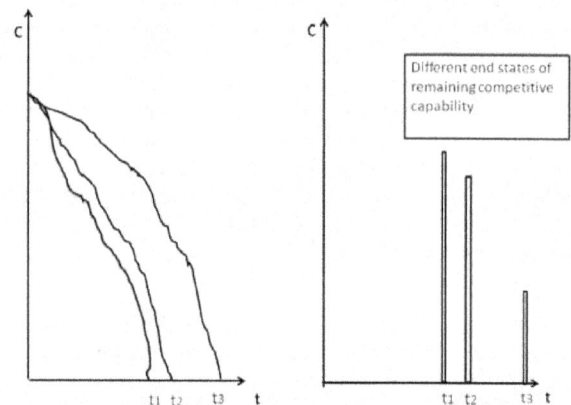

Fig. 2. Resolving the best tactics or strategy. By analysing more carefully the end state conditions one is able to resolve the best applied strategy. This figure emphasizes the remainig capability of competitor.

The remaining capabilities may be other own capabilities or owned by a competitor. A good strategy may then be a strategy maximizing own profit and minimizing the conflicting capabilities at the same time. On the other hand, some of the remaining capabilities may belong to the same side i.e. to ourselves or to a party sharing the same interests. Nonetheless, analyzing the end states of chosen strategies systematically gives definitely more information of the goodness of different strategies.

A schematic plot of the values of competiors capabililty of three different capability paths is shown in the Figure (2). The paths in the figure show the capability time evolution after the last decision-making situation, where three possible selections were available. Corresponding time evolutions of the capability after the last decision are shown in the figure. Each of the chosen strategy leads to an end state provided at capability's duration times t_1, t_2 and t_3. In this example, a further analysis reveals how the strategy 3 also minimizes the end state of the competitor's capability. In other words, if the competitor has conflicting interests and represents an opposing side, the strategy 3 was the most suitable strategy out of these three.

C. Time derivative of a capability for analysis of sudden changes

Since the capability depends on time and on the other capabilities as well as on their usage, the experimentally resolved behavior may be unexpected and complicated. It is difficult to find and analyze important occasions if only the capability curves are viewed as themselves. Human eye or mind is not a particularly effective tool for observing sudden changes or determining the steepness of a sudden change and not comparing the steepness of all curves with each other. In order to make the analysis more systematic and objective, one may take a time derivative of the capability function as it represents changes. The motivation for using time derivative is to observe sudden changes more easily as well as their steepness.

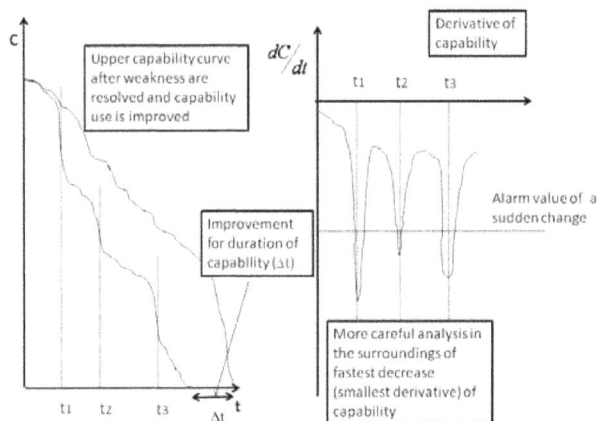

Fig. 3. By analyzing the time derivative of a capability one may observe more effectively interesting time intervals e.g. where the capability vanishes fastest. Such a reason might be e.g. ending of another capability or resource. One may then concentrate to study the corresponding time intervals and resolve the reasons for the sudden changes. In some cases one is able to avoid sudden changes by removing their cause(s). As an outcome, the duration of a capability may be extended or more profit can be achieved.

When the time derivative is taken one observes the values

of the sudden changes as peaks (or "wells" to be more specific). The value of the bottom of the well tells directly the steepness of the curve and therefore, the answer of a question "how fast the capability then decreases?". As a direct application, one may set "alarm"-values giving an alarm of a sudden change. The time derivative helps to find those particular time intervals for trying to resolve the leading cause(s) for the sudden change(s). For example, the sudden drop may occur at the instant vicinity of a vanishment of another capability or a resource. After resolving the leading cause one may analyze if the cause (e.g. vanishment of a resource) may be delayed. As a result, with the time derivative, one has better changes on delaying the vanishment (improving the duration) of a capability under interest.

In the Figure (3) there is a schematic example of an analysis of a capability curve. There are three sudden drops in the capability curve and they are seen as wells in the time derivative curve. After analyzing the surroundings of time t_1, t_2 and t_3 one may resolve the leading cause for the sudden changes. If one is able to delay the cause of one or more of the sudden drops one may be able to delay the total vanishment of the capability by Δt.

III. CONCLUSIONS AND DISCUSSION

We have considered sequential decision making process for understanding and visualizing the value of capability due to the decisions. We have introduced a model for comparing differential sequential decisions in terms of capability values in the presence of competing capabilities and using principles. The model can be applied for presenting duration of capability as well as analysis of profits obtained in the capability use. We called the strategies forming the decisions to capability paths as their graph visualize the values of capability with respect to time. In that manner, our model can be used to compare different tactics or strategies, i.e. path of decisions, and to find out which path leads to maximum duration of a capability. The time dependency of a capability is a very complex function and depends at least time, other capabilities, and capabilities interactions.

Since the capability is used in order to achieve profits or strategic goals, the capability should not vanish at too early stage. By analyzing the decision points (and the corresponding paths) of capability usage one is able to resolve the strategy leading to the maximum duration. The strategy or decisions leading to the maximum duration of the capability provides information of how long the capability can be available. However, the profits that the organization achieves in using the capability may be more interesting at least together within the information of capability duration.

This was further studied by introducing a procedure for analyzing the profits and remaining capabilities at the time instants where considered capability vanishes i.e. at the duration times corresponding to different strategies. In short, a mapping from duration times to the profits and remaining capability values. If the strategy leading to the longest duration of a capability leads also to the most profit or less capability of the competitor (and/or more capability of the other own capabilities), it is a strong indication of the chosen strategy being the most suitable one. Finally, we proposed to use time derivative of the capability function for more careful analysis of the duration of a capability. The fastest changes (decreases) in capability can be easily observed by time derivative minimum values and more careful study within surrounding intervals of minimum derivative values may be useful in finding out the leading reason(s) for the sudden drops. If the reason(s) can be identified, reduced, delayed or even removed, the duration of the considered capability may be improved by delaying the vanishment.

In applying our findings it is challenging to determine time evolution curves after decisions. However, there exists cases where capability decreases according to known rules. For instance in manufacturing, capability can be understood in terms of existing resources and the decision of the capability usage would concern on the capacity utilization. In that case, the visualization would emphasize e.g. the duration of the processed material.

The introduced model deals with capabilities duration in competitive environment. As the model is schematic, further studies can be conducted with respect to presented ideas. In particular, the question of capabilities usage and obtained profit would improve the understanding causal relations of own decisions.

REFERENCES

[1] D. Ulrich and M. Wiersema, "Gaining strategic and organizational capability in a decade of turbulence," *Academy of Management Executive*, vol. 3, no. 2, pp. 115–122, 1989.

[2] D. Ulrich and D. Lake, "Organizational capability: creating competitive advantage," *Academy of Management Executive*, vol. 5, no. 1, pp. 77–92, 1991.

[3] S. Y. Boyce, "Using intellectual capital and organizational capability to enhance strategic implementation for pharmaceutical firms," *Journal of Business and Public Affairs*, vol. 1, pp. ISBN 1934–7219, 2007.

[4] D. Teece, G. Pisano, and A. Shuen, "Dynamic capabilities and strategic management," *Strategic Management Journal*, vol. 18, pp. 509–533, 1997.

[5] K. Eisenhardt and J. Martin, "Dynamic capabilities: what are they?" *Strategic Management Journal*, vol. 21, pp. 1105–1121, 2000.

[6] C. E. Helfat, S. Finkelstein, W. Mitchell, M. Peteraf, H. Singh, D. Teece, and S. Winter, *Dynamic Capabilities: Understanding Strategic Change in Organizations*, C. E. Helfat, Ed. Blackwell Publishing, 2007.

[7] C. E. Helfat and M. A. Peteraf, "Understanding dynamic capabilities: progress along a developmental path," *Strategic Organization*, vol. 7, no. 1, pp. 91–102, 2011.

[8] A. Hinge, *Australian Defence Preparedness: Principles, Problems and Prospects: Introducing Repertoire of Missions (ROMINS) a Practical Path to Australian Defence Preparedness*, A. Hinge, Ed. Australian Defence Studies Centre, 2000.

[9] A. J. Tellis, J. Bially, C. Layne, and M. McPherson, *Measuring National Power in the Postindustrial Age*, A. J. Tellis, Ed. Santa Monica, CA: RAND Corporation, 2000.

[10] B. Bakken, "Handbook on long term defence planning," NATO Research Technology Organization, Tech. Rep., 2003.

[11] R. Martens and M. Rempel, "High-level methologies to evaluate naval task groups," *Naval Engineers Journal*, vol. 123, pp. 67–80, 2011.

[12] E. J. Arnold, "The use of military power in pursuit of national interests," *Parameters*, vol. Spring, pp. 4–12, 1994.

[13] J. H. Matlary and M. Petersson, *NATO's European Allies: Military Capability and Political Will*, J. H. Matlary and M. Petersson, Eds. Palgrave Macmillan, 2013.

[14] J. Kurth, "The iraq war and humanitarian intervention," *Global Dialogue*, vol. 7, no. 1-2, p. Winter/Spring, 2005.

[15] T. B. Seybolt, *Humanitarian Military Intervention: the Conditions of Success and Failure*, T. B. Seybolt, Ed. Oxford University Press, 2007.

Utilizing sensitivity analysis and data filtering processes to support overall optimizing and decision making models in military tactics

Jari Sormunen, Lt.Col.
National Defence University,
Finnish Defence Forces, Santahamina
Finland

Harri Eskelinen, D.Sc
Lappeenranta University of Technology
PL 20, 53850 Lappeenranta
Finland
harri.eskelinen@lut.fi

Abstract— **The goal of this research is to produce new information for developing the models of operational analysis and battle simulation applications used in Finnish Defence Forces. System models of military tactics are tools used to describe and study the relationships, contexts and effects of qualitative and abstract aspects. Operational research is utilized in the designing of mathematical optimizing models to find solutions for decision making in military tactics. Sensitivity analysis is applied to clarify 1) how the changes of the initial data collected from the battlefield affect the results of the mathematical analyzing method or of the simulating model of tactics, and 2) how these results later affect the leader's decision making. Two practical cases are discussed. The first example discusses the sensitivity aspects of stochastic modeling, and the second example describes the importance of sensitivity analysis of data collection during field exercises. In addition to sensitivity analysis, it is necessary to find a method for guaranteeing the content, quality and relevance of the observations collected from the battlefield. These three elements play a key role in the overall optimizing model. In addition, this paper also describes specialized filtering processes for filtering the data and observations from the battlefield and for making it possible for one to appropriately quantify observations for the optimizing procedure.**

Keywords—operational analysis; military tactics; sensitivity analysis; data filtering; optimizing; decision making

I. INTRODUCTION

In this paper, sensitivity aspects and optimizing viewpoints of the analysis model for tactical phenomena are discussed. This research is based on co-operation between Finnish Defence Forces and Lappeenranta University of technology from 2007 to 2012. The background research was carried out from 2003 to 2007, during which time 59 attack exercises were analyzed. The results of the background research are reported by Sormunen and Eskelinen in [1] and by Eskelinen and Sormunen in [2].

The main goal of this research is to produce new information and data for developing both models of operational analysis and battle simulation applications used in Finnish Defence Forces.

To reach this goal, this research clarifies how the tactical aspects examined affect the outcome of the attack exercise, from the viewpoint of unit performance optimization. The studied variables were divided into the following five main categories: tactics, situational awareness, loading, human factors and background aspects [2].

Our secondary goal is to develop methods for tactical analysis that make it possible to construct an exhaustive optimizing model for tactical decision making. The following questions were taken into consideration [2]:

- How does one measure and observe the tactical variables and other aspects, and what are the appropriate gauges?

- How does one analyze the effects related to battle outcome?

- How does one integrate qualitative and quantitative aspects to support tactical decision making and overall optimizing of the use of unit performance?

- How does one illustrate and interpret the results of the overall optimizing procedure?

II. SELECTED METHODOLOGY

There are several scientific approaches which could be applied to produce an exhaustive tactical overall optimizing model. One classic way of illustrating the functionality of different types of systems is the application of differential equations. In military tactics, differential equations could, for example, be used to illustrate the balance of dynamics between two fighting sides or to present the recovering process of the unit after evacuation actions because of the various casualty ratios involved. However, the system models could also be based on information theory, cybernetics, game theory, decision making theories, networking theories, stochastic models and operational research.

For the purposes of this paper, system models of military tactics are regarded as tools for describing and studying relationships, contexts and effects of qualitative and abstract aspects. On the other hand, operational research is utilized for designing mathematical optimizing models to find the best possible decision-making solutions in military tactics [Eskelinen and Karsikas, 3, p. 153].

The authors of this paper wish to emphasize the viewpoint that system theory may be regarded as a tool for understanding and illustrating the different functional parts of a system and how these different parts work together [Mielonen, 4]. The main point, however, is not to divide the system into its discrete parts, but rather to try to understand how the parts affect the unity of the system. Therefore, in this research, all the tactical aspects of the military unit (an infantry company) are discussed simultaneously, and the relationships and interactions thereof are studied as a whole. Finally, in this research, the focus is on the integrated overall result of tactical analysis, which is based on both the analytical approaches and system theory. This means that at times the discussion in this paper deals chiefly with the overlapping boundary areas of system theory and operational research.

III. IMPORTANCE OF SENSITIVITY ANALYSIS TO SUPPORT TACTICAL DECISION MAKING

Sensitivity analysis is applied to clarify 1) how possible changes in the initial data collected from the battlefield affect the results of either the mathematical analyzing or simulating model of tactics, and 2) how these results later affect the leader's decision making. Therefore, the principal model of sensitivity analysis consists of three main parts: sensitivity of the initial data collection, sensitivity of the mathematical analyzing method or simulation model, and sensitivity of decision making (see Fig. 1).

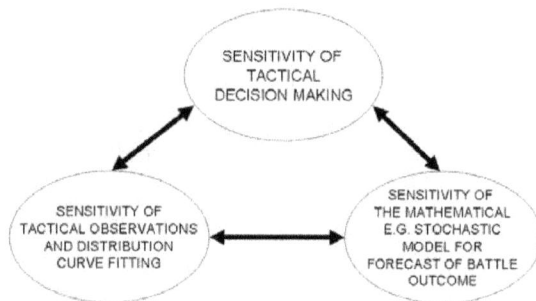

Fig. 1. Main parts of sensitivity analysis.

It is important to recognize, as illustrated in Fig. 1 by the two-way arrows, that the main parts of sensitivity interact with each other [3]. This means that on the one hand, the sensitivity aspect of decision making guides the initial data collection; on the other hand, the mathematical model forms the justification for the decision making. There are thus several tools used to estimate that enough data has been collected from the battlefield and that the accuracy of that data is high enough. In the analyzing and simulation models, there might be some uncertainty or instability, e.g. when one applies different distribution curve fits for forecasting the outcome of the battle. The model might yield wrong answers even though the initial data has been correctly collected. This is also the turning point of the sensitivity analysis. If the mathematical model produces distorted results, then the decision making is misled, too. The feedback from decision making towards the mathematical

models and initial data collection is needed to avoid unnecessary work if only general guidelines are expected as the result of decision making in lieu of an in-depth analysis of the aspects inside the decision making process. Rough decision making only requires tentative data collection and perhaps a simple distribution analysis based on assumed normal distribution variables.

Modern tactical simulations and mathematical analysis are based on stochastic modeling, which increases the importance of the appropriate sensitivity analysis [3]. This means that the models utilizing probabilistic analysis or randomized variables might turn into merely subjective war game models instead of scientific models used to estimate battle outcome. This will occur if the conditional clauses within the mathematical models utilize assumptions based on inaccurate sensitivity analysis. Indeed, this would approach manipulation of the results! Only careful sensitivity analysis ensures that the advanced mathematical models for supporting tactical decision making work properly.

Both quantitative and qualitative data are utilized in tactical research to form an overall picture of tactical phenomena from the battlefield. Sensitivity analysis of qualitative analysis is even more challenging than the previously described three-part method of quantitative analysis. The key to handling the problem is to ascertain the saturation limit, which describes when enough qualitative data has been collected. This basically means that the saturation point has been reached [3] when new qualitative data does not contribute anything additional to the explanation of the tactical phenomenon under research. However, it is important to remember that in qualitative analysis, we are not looking for the justification to generalize the results in a statistical sense.

Let us next discuss two practical cases where sensitivity analysis has played a key role in ensuring the quality of the research results in military tactics. The first example discusses the sensitivity aspects of stochastic modeling, and the second example describes the importance of sensitivity analysis of data collection during an attack field exercise.

A. Case 1: Sensitivity analysis and stochastic models

Some recent computational models for combat simulations used in Finnish Defence Forces are presented by Esa Lappi in [5 p. 14]. One goal of these studies has been to create tools for the optimization of the use of combat resources [5 p. 14]. The main aim of Lappi's study was to create methods for combat simulation tools for brigade-level land warfare scenarios. This work combines methods of mathematical modeling, stochastic processes and probabilistic risk analysis, all applied to battle simulation models suited to the Finnish combat environment. The results are needed for military planning, acquisition processes and cost-effectiveness calculations [5 p. 14]. Therefore, we can use the simulation model presented by Lappi as a case example for evaluating the importance of sensitivity analysis.

One section of the simulation model describes the evacuation process of a wounded soldier at the platoon level. In this part of the simulation model, the operator provides the average evacuation time and the standard deviation for the connection. The phenomenon can be represented as a Markov chain, whereby the patients move from one state to the next through the evacuation chain. This contains both the state transitions for treatment and the state changes without treatment. Given the average time μ and the standard deviation σ of the time, the number of states can be calculated as $N = (\mu/\sigma)^2$. However, the number of states is set to be at least 2 and at most μ/σ [Lappi et al 6].

Several measurements have been performed to validate this model, some of which are discussed in detail by Lappi et al in [7]. Part of the experiment was recreated in Sandis software, which produced very similar results. During the tests, the casualty rate was determined to be 35.4%, while the Sandis software result was 37%. If the probability of a hit is 35% with 100 targets, the standard deviation $\sigma = (npq)^{1/2} \approx 5$, so $35\pm5\%$ is the ideal accuracy for the hit probability in a single test [7]. But what happens when the expected distribution does not follow normal distribution function is that the given estimation of the evacuation process might be totally misleading.

B. CASE 2: Sensitivity analysis and field exercises

Our second case example focuses on evaluating the importance of the sensitivity analysis conducted for data collection of a soldier's feeling of having the initiative during an attack exercise. In the literature dealing with tactics [Rekkedahl, 8, p. 79-80], in the Finnish Defence Forces regulations and field manuals [9] and in recent tactical research [Huttunen, 10, p.125-128], initiative is regarded to have an essential impact on the success of a battle. In addition to these findings from the literature, General Tynkkynen, for example, also asks, should the execution model of the areal defence be changed because based on his viewpoint Finland needs an execution model that is more initiative than the one at present [Tynkkynen, 11, p. 92]. The literature research shows [2] that the most significant differences between successful and unsuccessful attacks are the ability to deal with the following aspects based on company leaders' answers in field exercises:

- Initiative
- Adversary's ability to effect
- Physical task load
- Timely task load
- Mental task load

The human factors seem to be important for the success. In the background research, soldiers' feelings of having initiative were measured by questionnaires and interviews. The results showed that the distributions of the results, which presented the soldier's on both fighting sides having feelings of initiative, partly overlapped [12]. It is therefore interesting to research whether the quantified results from the qualitative analysis of the background research support this finding of overlapping distributions and what new aspects sensitivity and distribution analyses could bring to the discussion.

The practical measurement system is developed in NASA-TLX-research as described by Sandra et al in [13, p. 139-184]. NASA has utilized the same type of measurement and data collection system for its astronauts to find out how to maintain astronaut performance in a weightless environment. The analogous similarity to the battlefield environment stems from the fact that in the battlefield also, the question is how to maintain human performance in an environment which departs from the normal. In principle, this type of measurement and data collection system also works in the battle exercise if the soldiers evaluate their own feelings [Härkönen et al, 14]. In the background research, the soldiers of both fighting sides (distinguished here as blue or red side) were asked to evaluate their own feeling of having the initiative at each measurement level of an attack, by marking the point on a 100 millimeter-long line indicating the strength of their feeling (see Fig. 2). With this form of measurement, quantified data was collected in a continuous range. This way of producing quantified and scaled data made it possible to carry out the mathematical analysis dealing with the feeling of having the initiative.

$$0 \longrightarrow\!\!\!\!\!\!\!| \longrightarrow 100 \text{ mm}$$

Fig. 2. Measuring and quantifying the data dealing with the soldiers' feeling of having initiative.

To verify the hypotheses, it is necessary to establish which fighting side has the initiative at each moment of the attack and how strong the initiative is. This is done by calculating the size of the overlapping area of the two distributions describing the variation of one's own forces' feeling and the adversary's feeling of having the initiative. The probability that both sides feel they have the initiative is calculated. The distribution analyses dealing with the feeling of having the initiative are carried out in the stepwise analysis illustrated in Fig. 3.

Firstly, mean values (μ_b-μ_r) and standard deviations (σ_b, σ_r) of feeling of having initiative at each measurement level are calculated for both fighting sides. Secondly, the size of the overlapping area may be roughly estimated e.g. by applying the calculated nomogram [Linacre, 15, p. 487-488]. Thirdly, after the previous nomogram has revealed the overlapping areas, the probability of the different opinions between the blue and red side of having the initiative is calculated by the equation (1).

$$z_p = \left| -\frac{\mu_b - \mu_r}{\sqrt{\sigma_b^2 + \sigma_s^2}} \right| \tag{1}$$

where

z_p = Normal distribution coefficient

μ_b = Mean value (own side)

μ_r = Mean value (adversary side)

σ_b = Standard deviation (own side)

σ_r = Standard deviation (adversary side).

The dependency of having the initiative on the fulfillment of the battle task and on the number of suffered casualties is defined by establishing the probability values of the initiative and the corresponding casualty ratio at each measurement level

At first the values of mean and standard deviation are calculated for feeling of having initiative of both fighting sides based on measured data in range 0...100 mm.

The probability of the overlapping area is calculated according to the given equation. The size of this area describes, how strong both fighting sides feel, that they have the initiative

In cases where, distributions of initiative overlap remarkably, a rough estimation of overlapping percent (the coloured area between distributions) is made by utilizing the principal set of estimation curves.

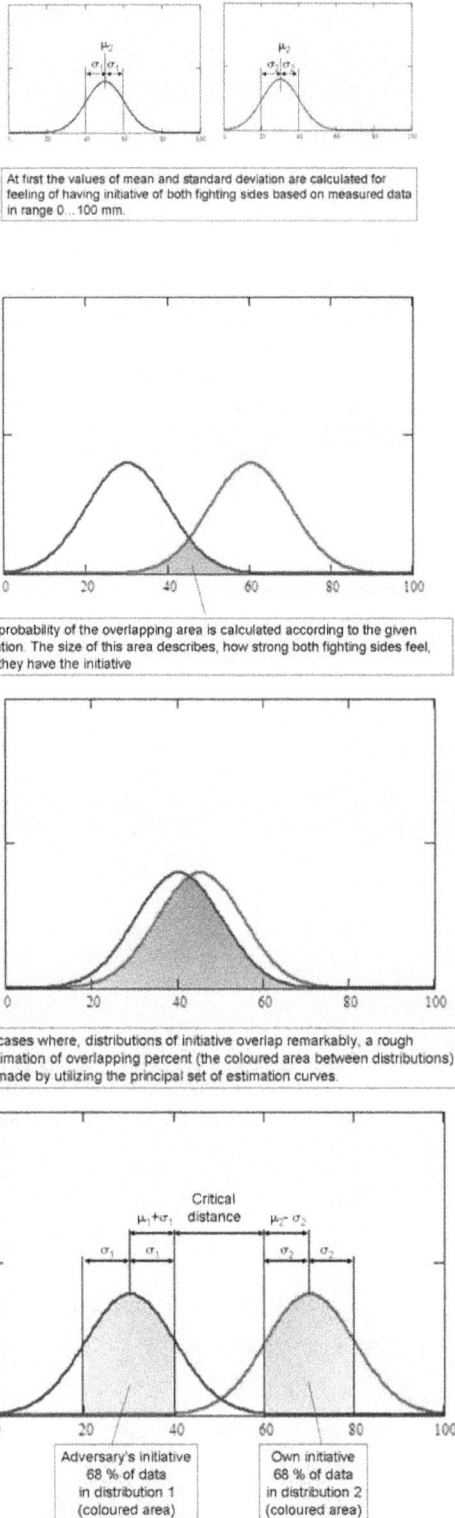

Fig. 3. Stepwise distribution analysis dealing with the feeling of having initiative.

The next step is to calculate if the advantageous initiative indicates a positive progress of the casualty ratio with a high correlation. What would happen if the distributions do not follow the normal distribution curve is that the tactical assumptions would be misleading.

These two cases demonstrate that an overall optimizing model for evaluating military tactics with probabilities should be developed to also recognize the distributions which do not follow the expected normal distribution curve accurately enough. Kapur [16] has presented either derived equations or readily calculated resulting tables for analyzing situations, whereas sensitivity analysis requires the calculation of the overlapping area of two normal distributions, two log-normal distributions, two Poisson's distributions, normal and Poisson-distribution, two Gamma-distributions, normal and Weibull distribution and two Weibull distributions.

Weibull distribution is mathematically quite flexible, because its curve can be fitted satisfactorily to follow the various types of results. Therefore, it is interesting to see if it could be the solution for both previous case examples for improving the sensitivity of the tactical analysis. In [16] there are numerical values in the tables which can be used to solve the probability of the area where the normal distribution overlaps with the Weibull distribution (see Fig. 4).

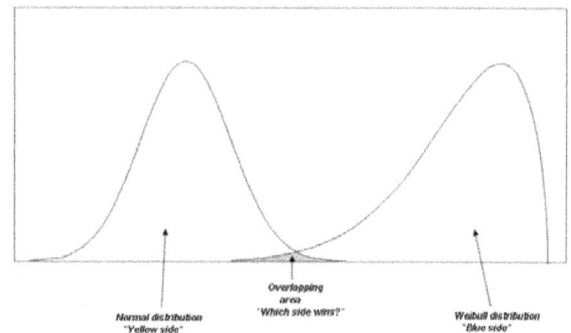

Fig. 4. Overlapping area of normal and Weibull distribution.

If the parameters λ and η of the Weibull distribution and the parameters σ_s and μ_s of the normal distribution are known, and if these distributions overlap, two relationships can be written [Kapur and Lamberson, 11] as shown in (2) and (3):

$$C = \frac{\lambda - \eta}{\sigma_S} \tag{2}$$

$$A = \frac{\eta - \mu_S}{\sigma_S} \tag{3}$$

By substituting here the values of the cutting and scaling parameters of the Weibull distribution and the ratios calculated from the mean values and standard deviations, it is possible to solve the values of coefficients A and C expressed in (2) and (3) and, furthermore, to read the probability of the overlapping area by utilizing the calculated tables [16]. It is likely that when one analyzes the battle exercises (in the form of data sets describing some tactical aspects of both fighting sides), the distributions of some variables may not follow normal distributions [3, p- 123-127]. To be able to forecast the battle outcome, the use of the Weibull distribution described here would be justified.

According to our results, it is possible to utilize the NASA-TLX index to try to quantify the differences between the emotional or human factors between the fighting sides, such as "the feeling of having initiative", but it is possible that these human factors indicate the outcome of the battle incorrectly. In some cases, the leader's positive feeling of having the initiative might have been based on e.g. misleading situational awareness or unexpected changes of the tactical situation, and then, during the battle that positive feeling has collapsed. However, it is possible that due to some other successful factors the leader is still able to lead his unit to victory. Although the effort to try to quantify the feeling of having the initiative might give some new information from the battle field, it should be noticed that the task is at least challenging and special attention is needed to verify the effects.

IV. DATA FILTERING FROM BATTLEFIELD TO CONSTRUCT THE OVERALL OPTIMIZING MODEL

In addition to sensitivity analysis, finding a method to guarantee the content, quality and relevance of the observations collected from the battlefield is necessary. These three elements play a key role in the overall optimizing model (see Fig. 5). In this research, we describe the specialized filtering process utilized to filter the data and observations from the battlefield, to make it possible to quantify these appropriately for the optimizing procedure.

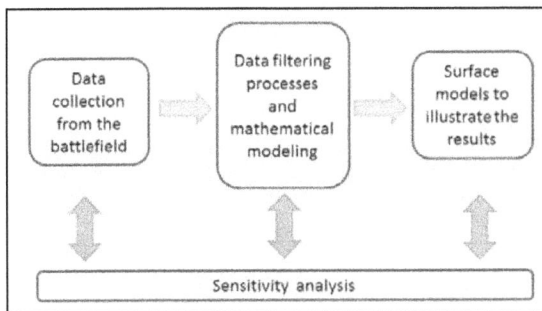

Fig. 5. Basic elements of the overall optimizing model.

The development and the justifications of the data filtering processes is based on the material, which was collected in the background research and which is documented by Finnish Defence Forces (FDF). This material consists of the result database [17, p. 1-2] including results of 59 attack exercises, tactical maps and event catalogues of these attacks and 103 written reports from data collectors (professional officers on duty at FDF). The filtering processes are used to organize and handle the data sets of 118 measured variables and 130 observed factors during each attack. [18] The data included 11912 recorded fighter attack actions, which were collected during 34 measurement days.

The filtering process is carried out by applying five different analyses: result-dependent, battle space-impulse, time-dependent, decision making-oriented, and sequential effect-chain analysis. In result-dependent analysis, the value of

each observation is derived into a numerical value according to the result of the required action. In battle space-impulse analysis, if the attack is successful, different impulses lead to initiative, dynamic, creative and dominant functions instead of to post-reactive corrections affecting the values of entered numerical data. When time-dependent analysis is carried out, phenomena are observed based on two time variants, which affect the entered data values. Decision making-oriented analysis handles the data pertaining to the leader's decision making and his tactical solutions. Within the sequential effect-chain analysis, tactical analysis is employed in an attempt to find the most important connections between different tactical aspects.

The aim of the overall optimizing procedure is to integrate several different variables and affecting factors into one tactical overall model 1) according to the company leaders' decisions, solutions, and orders; 2) according to the unit's actions; and 3) according to the events in the battle space. Special filtering processes are needed to derive mostly quantitative observations into numerical values describing tactical phenomena.

The filtering processes are divided into the following five sub-analyses categories in order of importance:

- Result-dependent analysis
- Battle space-impulse analysis
- Time-dependent analysis
- Decision making-oriented analysis
- Sequential effect-chain analysis.

The most important sub-analysis is thus result-dependent analysis. A simplified chart of this analysis is presented in Fig. 6. The flexibility between decision making and the desired successful course of action is highlighted.

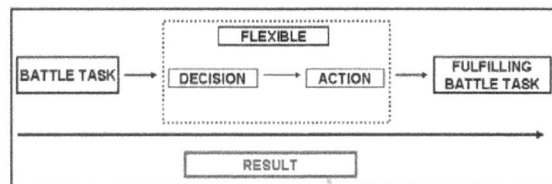

Fig. 6. In the result dependent analysis the quantity is determined by the result of the selected action. However, flexibility in the evaluation of interaction between decision and action is allowed.

The second most important filtering principle in data collection is battle space-impulse analysis. This highlights the aspect that any impulse at any time during the battle should lead to initiative, dynamic, creative and dominant functions in the company for dealing with the current situation and for winning the battle or for supporting battle success (see Fig. 7).

It is important to acknowledge that this analysis is related neither to theoretical time-axes nor to the principal distance-axes of the infantry company's attack. In this analysis, the observations are made by examining changes in the company's functions in accordance with each event in the battle space.

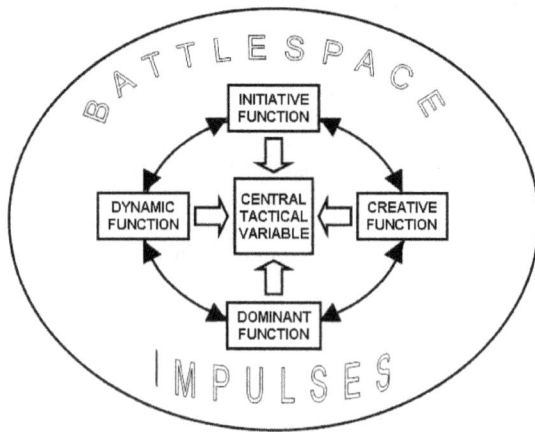

Fig. 7. The Battlespace impulse analysis. In a successful attack, impulses should lead to initiative, dynamic, creative and dominant functions instead of post-reactive corrections.

The third most important filtering process is time-dependent analysis. In the principal progression of an attack exercise, the time boundaries are variable, but the time axes of a real attack are, of course, continuous. This is illustrated in Fig. 8. Time measurement starts from readiness and ends with continuity. In addition to this, observations are made in the three-dimensional coordination of effect, decision and time. This addition makes it possible to quantify the speed of each effect and decision and relate these to the outcome-dominated overall tactical evaluation. Furthermore, this enables one to calculate the time needed to reach the maximum speed of the selected course of actions (acceleration).

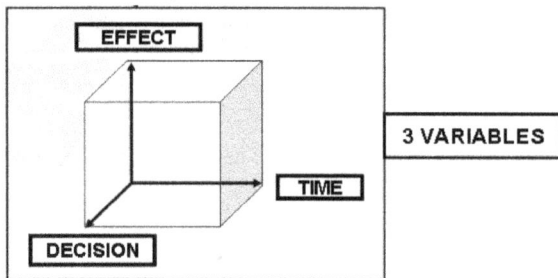

Fig. 8. The time dependent analysis.

Fourth in order of importance, decision making-oriented analysis follows the model of the data-flow process, which is time dependent. In this case, the input of the process consists of affecting variables and background factors which integrate with information connected to situational awareness. The human decision making process itself is not essential, but of main interest is the discovery of the connections between the decisions made, the presumable results and the time required for the tactical decision making.

The decision making process is seen as initiative, concurrent and active action. The leaders' and soldiers' own active and personal situation evaluation influences the essential factor in this case.

The output of this process model yields knowledge that has already been analyzed of the responses and effects due to the decisions made. Furthermore, the result of the process is summarized and observed in the company's tactical solutions and actions connected to the leaders' decisions. This process model is illustrated in Fig. 9.

On the one hand, any action in most organizations starts with a command given by the boss, which is assumed to lead to a desired effect through maneuvers by staff. Every modern organization strives to maintain its performance as well. On the other hand, the established cornerstones of decision making are background knowledge, activeness, simplicity, concentration of effort and backdoor thinking. In our research, these two viewpoints are modified for the tactical analysis of an infantry company's attack. The viewpoints form the basic chains within the sequential effect-chain analysis.

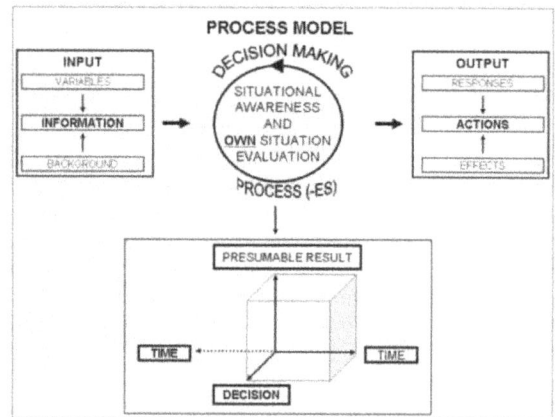

Fig. 9. Process model of the decision making oriented analysis.

The development of both time-dependent analysis and decision making-oriented analysis originate from ideas of applying principles dealing with experiencing time in leading systems as noted by Kuusisto and Helokunnas [19, p. 415-419]. According to these principles, the aim should be remembered when one studies time in the context of leading situations. When these principles are applied to the area of tactical analysis, another important viewpoint deals with how to combine the theories of time with three types of information, which are 1) information about the past, 2) perceptions about the current moment and 3) intuition about the future [Kuusisto and Kuusisto, 20, p. 78-81]. The cubic model in Fig. 9 shows how this viewpoint has been applied in decision making-oriented analysis.

In sequential effect-chain analysis, specific tactical variables establish the key chain connection with respect to the battle outcome. It is possible for one to analyze and predict the success of a tactical decision and the action connected to it even if using only the key chains. In these cases, the linkages between the tactical variables are in a central position. One typical key chain consists of three tactical variables: 1) freedom of action, 2) execution capability and 3) utilization of success.

V. RESULTS AND DISCUSSION

Stepwise sensitivity analysis plays a key role in the construction of the overall optimizing model of tactical analysis. What is likely easiest to accept is the need for an appropriate curve fitting the distribution of measured variables from the battlefield. It is too often and too easily assumed that normal distributions of measured tactical variables are accurate enough to be included in the mathematical (e.g. stochastic) models to form the basis for estimating or forecasting the outcome of a battle. Our results have shown that if the appropriate distribution function is found, the reliability of the optimizing model can be improved without any other essential changes made to the overall model.

The most important result of this research is the development of filtering processes for handling the qualitative observations from the battlefield and for forming the basis for qualifying them for use in the mathematical optimizing models. The conclusions that are drawn from the five sub-analyses processes presented in this paper could be summarized as follows. The purpose of result-dependent analysis is to verify if the battle task has been fulfilled. Battle space-impulse analysis confirms whether the course of the battle was victorious or not. The point of time-dependent analysis is to check the success of the battle timing. The decision making-oriented analysis confirms the justification of the main decisions made relevant to the battle situation. The sequential effect-chain analysis indicates and ascertains the key tactical variables and the importance of these in the battle.

Finally, as a concrete tactical result taken from the filtering process conclusions, it is possible to extract and call attention to certain tactical key points of the battle:

- Was the battle task fulfilled and was the battle victorious?

- At what cost was the outcome of the battle achieved?

- Was the timing related to the tactical decisions and tactical actions correct?

- Were the decisions and tactical actions actively justified or merely conducted "like before"?

- What are the key variables and the culmination points of the attack?

By way of conclusion, one way of illustrating the results of the tactical overall optimizing procedure is to utilize a surface model (see Fig. 10).The set of coordinates is formed of the tactical principles and tactical elements, and of the corresponding quantified values of the tactical observations dealing with those principles and elements. The sequences of the rows and columns inside the CMEP-matrix, were established by utilizing the field manuals, regulations and other relevant literature. Further on, the final construction of the model presented in Fig. 10 is bases the qualitative analysis according to Miles and Huberman [21, p. 10-12], during which about 12,000 observations from the background research were analyzed. These observations were produced by professional officers of the Finnish Defence Forces. In the Finnish Defence Force's Field Manual, the central tactical principles are described with the word, "consciously, actively, simply, concentrated and continuously". According to the field manual

the company leader leans on these general tactical principles when he is leading the battle and the compliance with these principles makes it possible to succeed in the battle. According to the Field Manual the tactical elements are command and control, manoeuvre, effect and performance maintenance. The tactical principles become apparent in the interaction with the tactical elements of the battle [9].

These kinds of surfaces are generated by applying measured data into a Beziér-surface, which makes it possible to add mathematical comparison and pattern recognition to the decision making process. As illustrated by the different colored layers in Fig. 10, different types of coefficients are easy to add for purposes of adjusting the height of the surface for example to describe the difficulty level of the tactical task or to scale the final grading of the measured attack.

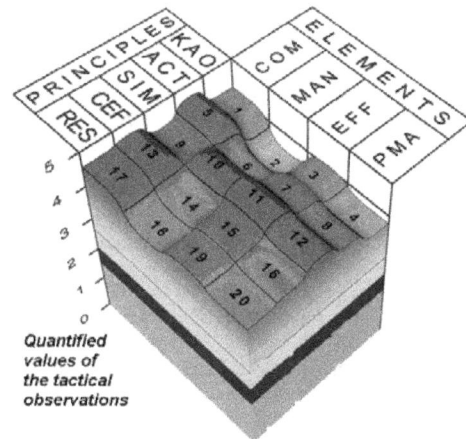

Fig. 10. Surface illustration of the tactical overall optimizing procedure.

Both during the background research and during the ongoing research at Finnish Defence Forces and Lappeenranta University of Technology it has been seen that there are different types of uncertainties in the battlefield, which should be recognized. In this paper only one type of uncertaintity has been examined in details. This paper has dealt mostly with the possibilities of utilizing appropriate distribution types to support tactical analysis and sensitivity aspects connected with that. However, uncertainties related to battle simulations can be divided into different groups, such as variability, precision, ignorance or vagueness. First kind of uncertainty (variability) is typically found with the data which deals with the so-called human factors (like soldiers' feeling of having initiative discussed in this paper).

It is also noticeable, that the equal values or magnitudes of collected data do not necessarily show identical states (precision). On the other hand, the sensitivity is connected with the establishment of the appropriate reference level of the precision. However, this level depends also on the environmental aspects including the effects of observer's own presence in the test set up and therefore the level is difficult to set at any exact height.

When the observer collects data from the battlefield, we must remember and recognize the problem of "ignorance": The observer may suffer from the lack of required knowledge to analyze the situation, he might just collect his beliefs of what has happened instead of "real" observations or in some cases he might even ignore some important evidence.

Vagueness describes the observation itself. The complex situation at the battlefield causes the problem that some "observations" might be unclear or uncountable.

Because we know, how complex the tactical analysis might be and, yet, clear decisions are needed to carry out the appropriate military manoeuvres, it is reasonable to utilize different types of tools to increase the reliability, validity and effectiveness of the tactical analysis. One option is to utilize sensitivity analysis and developed data filtering processes to support the overall optimizing and decision making models in military tactics.

REFERENCES

[1] J.Sormunen J. and H.Eskelinen, Mitattua taktiikkaa – Komppanian hyökkäyksen menestystekijät, Edita Prima Oy, Helsinki, 2010.

[2] H. Eskelinen and J.Sormunen, Kvantitatiiviset ja kvalitatiiviset tutkimusmetodit komppanian taisteluharjoituksen analysoinnissa, Tutkimusraportti 88, LUT,Teknillinen tiedekunta, Lappeenranta 2012.

[3] H.Eskelinen and S.Karsikas, Tutkimusmetodiikan perusteet - Tekniikan alan oppikirja, LUT Koulutus- ja kehittämiskeskus, Julkaisu 12, Lappeenranta 2012.

[4] S.Mielonen, Systeemiajattelu, http://www.mlab.uiah.fi/polut/Yhteiskunnalliset/tyokalu_systeemiajattel u.html, down loaded 30.7.2010.

[5] E. Lappi, Computational methods for tactical simulations, Maanpuolustuskorkeakoulu, Taktiikan laitos, 1/2012, Juvenes Print Oy, Helsinki 2012.

[6] E.Lappi, B.Åkesson, S.Mäki, S.Pajukanta and K.Stenius. "A Model for Simulating Medical Treatment and Evacuation of Battle Casualties", 2nd Nordic Military Analysis Symposium, Stockholm, Sweden, Nov.17-18, 2008.

[7] E.Lappi, O.Pottonen, S.Mäki, K.Jokinen, O-P.Saira, B.Åkesson and M.Vulli, "Simulating Indirect Fire – A Numerical Model and Validation through Field Tests", 2nd Nordic Military Analysis Symposium, Stockholm, Sweden, Nov, 17-18, 2008.

[8] N.M.Rekkedal, Nykyaikainen sotataito – sotilaallinen voima muutoksessa, Maanpuolustuskorkeakoulu, Taktiikan laitos, 2006.

[9] Kenttäohjesääntö, yleinen osa (KO YL), Prikaatin taisteluohje (PR TST-OHJE) and Komppanian taisteluohje (KOTO), Finnish Defence Forces, 2008.

[10] M.Huttunen, Monimutkainen taktiikka, MPKK, 2010.

[11] V.Tynkkynen, Taktisen ajattelun haasteet – Katoaako oma-leimaisuutemme, Sotataidon jäljillä – Taktinen ja viestitaktinen taito taistelukentällä, MPKK, 2004.

[12] National Defence University of Finland, MPKK AF15339, Jakaumatarkastelut, Finnish Defence Forces, 2009.

[13] G.Sandra, G and L.Staveland,"Development of NASA-TLX (Task Load Index): Results of Empirical and Theoretical Research, Human Mental Workload", Advances in Psychology 52, pp. 139-184, 1988.

[14] T.Härkönen, M.Vulli, H.Eskelinen and L.Oksama, "The Effect of Infantry Simulator Training on Combat Performance", 54th IMTA Conference, Dubrovnik, CROATIA,5.-9.11.2012.

[15] J.Linacre, "Overlapping Normal Distributions", Rasch Measurement Transactions, Vol.10:1, Spring, pp. 487-488, 1996.

[16] K.Kapur and, L.Lamberson, Reliability in Engineering Design, John Wiley&Sons Inc., USA 1977.

[17] PEMAAVOS R6524: Komppanian hyökkäyksen menestystekijät, toimeksianto, Finnish Defence Forces, 2003.

[18] MAASK AD5337 and MAASK AD5337: Tutkimuksen päättäminen ja esitys jatkotoimista, Finnish Defence Forces, 2007.

[19] R.Kuusisto and T.Helokunnas, "Experiencing time in leading systems", Proc. of 2003 IEEE International Engineering Management Conference, Albany, New York, USA, pp. 415-419,2003.

[20] R.Kuusisto and T.Kuusisto, "Strengthening Leading Situations via Time-divergent Communication Conducted" in J.Van Beveren, The E-Business Review, Volume III, the International Academy of E-Business, Nacogdoches, Texas, USA, pp.78-81,2003.

[21] M.B.Miles and M.A.Huberman, Qualitative Data Analysis, Second Edition, Sage Publications, USA, 1994.

A First Approach to the Optimization of Bogotá's TransMilenio BRT System

Francisco J. Peña, Antonio Jiménez and Alfonso Mateos

Decision Analysis and Statistics Group (DASG)

Departamento de Inteligencia Artificial, Universidad Politécnica de Madrid

Campus de Montegancedo S/N, Boadilla del Monte, 28660, Madrid, SPAIN

Email: francisco.pena.escobar@alumnos.upm.es, {ajimenez, amateos}@fi.upm.es

http://www.dia.fi.upm.es/grupos/dasg/index.htm

Abstract—Bus rapid transit (BRT) systems are massive transport systems with medium/high capacity, high quality service and low infrastructure and operating costs. TransMilenio is Bogotá's most important mass transportation system and one of the biggest BRT systems in the world, although it only has completed its third construction phase out of a total of eight. In this paper we review the proposals in the literature to optimize BRT system operation, with a special emphasis on TransMilenio, and propose a mathematical model that adapts elements of the above proposals and incorporates novel elements accounting for the features of TransMilenio system.

Keywords—Bus rapid transit systems, optimization, mathematical modeling.

I. INTRODUCTION

Bus rapid transit (BRT) systems are public transport systems with medium/high capacity, high quality service and low infrastructure and operating costs ([1]). They are considered to be a good affordable alternative for developing cities seeking to provide their citizens with a high-quality possible self-sustaining public transport alternative comparable with rail systems, but without the high costs and without taking cities to high levels of debt, leaving the possibility of investing the city funds in priority areas such as health or education.

BRT systems have a lot in common with rail systems, particularly performance and passenger service. The main difference is that operation and implementation costs are 4 to 20 times lower than the costs of a light rail system, and 10 to 100 times lower compared to a heavy rail and metro system ([1]).

They can operate of limited stop services (also called stop-skipping services), in which a bus service omits stops along certain routes. This has great advantages, such as the reduction of travel times due to fewer stops and the reduction of operator costs because they can meet the demand with fewer vehicles thanks to shorter bus cycles ([2]).

BRT systems are now operating in 149 cities, most of which have been built since 2000, and 84 more are planned around the world. TransMilenio is Bogotá's most important public transportation system and one of the biggest BRT systems in the world. New plans have been made to expand it due to its success, and similar systems have been constructed in other cities of Colombia.

There are very few proposals in the literature focused on optimizing the BRT system operation, mainly because they are relatively recent phenomena, and many of the currently operating BRT systems are far from reaching maximum capacity. To the best of our knowledge, there are not automatic proposals for route design. The closest to this is the model proposed in [2]) that evaluates and selects the best several routes.

In this paper we review of the proposals in the literature to optimize BRT system operation, with a special emphasis on TransMilenio, and propose a mathematical model that adapts elements of the above proposals and incorporates novel elements accounting for the features of TransMilenio system. Specifically, we introduce a new model for evaluating Trans-Milenio BRT system routes, given the trip demand in the form of an origin-destination matrix.

Section 2 introduces BRT systems and their main elements. Section 3 focuses on the TransMilenio system, Bogotá's most important public transportation system and one of the biggest BRT systems in the world. In Section 4, we review the different studies in the literature on the optimization of BRT systems and, specifically, on the TransMilenio system. In Section 5, we introduce a new mathematical model approach to the optimization of the TransMilenio system. Finally, some conclusions and future research are discussed in Section 6.

II. BUS RAPID TRANSIT SYSTEMS

A BRT system was defined in [1] as a system based on high quality buses, that provide fast and a comfortable urban mobility and with a favourable cost-benefit through the provision of segregated infrastructure of exclusive use, fast and frequent operations, and marketing and customer/user service excellence.

The first BRT system started operating in Curitiba in 1974, but until the decade of 1990 this type of system was seen as a public transportation system for small cities or as complementary systems of a metro network. Many experts considered that these systems were not able to reach a capacity beyond 12000 passengers per hour per direction (pphpd). This perception radically changed in 2000 with the creation of TransMilenio in the city of Bogotá (Colombia). Nowadays, TransMilenio transports nearly 500 million people yearly ([3]). It introduced a series of improvements that raised the capacity of BRT systems enormously to 45000 pphpd, and has inspired

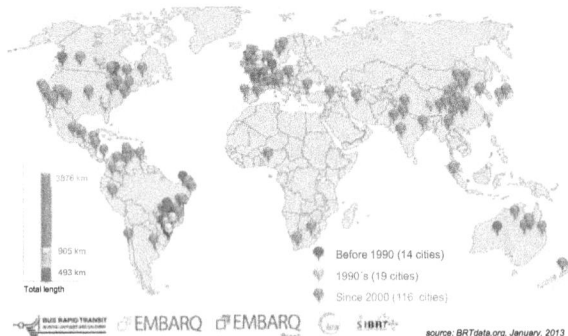

Fig. 1. Evolution of number BRT of cities and km per decade ([4])

many cities around the world to implement this type of systems ([1]).

Nowadays, there are 149 cities with BRT systems, and 84 more are planned. The majority of systems were built after the year 2000, which can be attributed to TransMilenio's success, as illustrated in Figure 1, which shows the evolution in the number of cities with BRT systems per decade and the respective number of kilometers.

A. Comparison with other mass transportation systems

Table I shows the price range for mass transportation systems based on a comparison of infrastructure costs real data ([1]).

Type of system	Cost per kilometer (US$ million/km)
BRT	0.5 - 15
Tram and light rail transit	13 - 40
Elevated systems	40 - 100
Underground metro	45 - 350

TABLE I. CAPITAL COSTS FOR DIFFERENT MASS TRANSPORTATION SYSTEMS

The infrastructure costs for BRT systems are clearly significantly lower than for any other rail-based transportation system. The city of Bangkok is a case in point. This city has an elevated rail system (SkyTrain) and an underground metro system (MRTA), a proposed BRT system (Smartway) and a proposed light rail train. The real costs per kilometer of the metro systems and elevated train were US$ 142.9 million and US$ 72.5 million. The projected costs per kilometer for the proposed light rail and BRT are US$ 25 million and US$ 2.34 million. This means that with a budget of US$ 1000 million they could build 7 km of underground train, 14 km of elevated train, 40 km of light rail train or 426 km of BRT system ([1]).

Unlike rail systems around the world, BRT systems are capable of operating without government subsidies. As a matter of fact, they are profitable, which is the reason why many governments delegate the operation to private companies. This is a great advantage, especially useful for developing cities, where governments have tight budgets and there is

nothing better than a self-sustaining mass transportation system thanks to which they can invest resources in other areas such as sewerage, education and health.

BRT systems can be planned and implemented in short time periods, which can be covered in one government term. The two most successful and complete BRT systems (Curitiba and Bogotá) were planned and implemented in a three year span.

Formerly it was thought that bus based services could operate within a range up to 6000 pphpd. If the demand was higher, a light rail based system should be considered, with capacity between 6000 and 12000 pphpd. A heavy metro system had to be considered for a higher demand, since its capacity ranges from 25000 to 80000 pphpd.

The arrival of BRT systems with a capacity range from 3000 to 45000 pphpd changed the situation. BRT systems turned into a real mass transportation alternative for big cities, and the myth that says that BRT system cannot compete with metro systems in terms of capacity was broken. As a matter of fact, it is not necessarily true that big cities need overflowing levels of capacity, an example is the London metro system, which has a capacity of 30000 pphpd, but thanks to its multiple parallel lines it has distributed corridors demand across the entire transportation network. An opposite case is Hong Kong's metro, whose capacity is 80000 pphpd and there is only one line from Kowloon and New Territories to Nathan Road. But the high level of demand is what makes this metro system profitable ([1]).

B. Main components of a BRT system

A BRT system has seven main components ([5],[6]): busways, stations, vehicles, fare collection, intelligent transportation systems, service and operation plans, and branding elements.

The *busways* or *corridors* are the main component of the BRT systems and it is where the vehicles circulate; they are like the rails of a metro system. They are also the most expensive and one of the most visible elements of the whole system. Therefore, they have a direct impact on the image and perception the users have of the system ([6]). The busways must be exclusive for the system buses. Furthermore, the busways must be located in the center and not at the side of the road ([1]).

The *stations* are the link between the passengers, the BRT system and other transportation systems. They are the element that has most influence on system image, and so, they must have comfortable facilities so that the passengers feel at ease. The stations must accommodate many more people than a bus stop, so they must have a wide infrastructure ([6]) since they are located in high demand busways. Besides, the stations must protect the users from climate conditions. The boarding platforms must be at the same height as the buses floor to ease and speed up passenger' access. There must be also large capacity header stations at the ends of each busway to integrate busways and feeder routes [1].

The *vehicles* are the system's element in which the passengers spend most of their time. They have a direct impact on speed, capacity, comfort and environment friendliness. They are the element that most non-users see, becoming one of the elements with most influence over the public's perception of the system ([6]).

Currently, there are three types of vehicles: articulated, bi-articulated and simple. *Articulated buses* have the capacity for 160 passengers and operate within the busways. *Bi-articulated buses* have the capacity between 240 and 270 passengers and operate within busways. *Simple buses* have the capacity for 90 passengers and in some cities they operate only in mixed traffic corridors as feeder routes. In other cities with low capacity BRT systems operate in busways.

Fare collection has a direct effect on capacity and the system's income. If fares are collected outside the bus, it decreases passenger waiting time through bus boarding efficiency. This is especially useful for bus routes that have high levels of demand ([6]). The fare may be collected outside the bus at the station entrance. Furthermore, fares should be integrated, that is, users should be able to transfer from one bus route to another, including feeder routes, without having to pay an extra fare [1].

The *intelligent transportation system* is a technological component that helps to improve overall system performance. It is a combination of different technologies to retrieve all kind of data about system operation, from the number of passengers that enter the system to the positioning of every vehicle using GPS, vehicle departure times, traffic conditions, the traffic lights, etc. The goal of this component is to collect and transform all the possible information into useful knowledge for operators, and ultimately benefits for the passengers.

The *service and operation plans* directly affect the user's perception of the system. A good plan allows to adjust to the levels of demand present along the busways. Frequencies should be high to reduce waiting times, and a good design will also help to reduce the number of passenger transfers. Furthermore, the route maps must be easy to understand for users [6]. The busway and feeder routes must be physically integrated, forming a network. Besides, the entry of other public transport operators must be restricted ([1]).

The BRT systems must have a distinctive *brand* image from other transportation systems. A good marketing strategy can position the brand and improve its image to attract more users. The BRT system should have a positive brand image.

III. TRANSMILENIO BRT SYSTEM

TransMilenio is Bogotá's most important mass transportation system and one of the biggest BRT systems in the world. It is based on the Curitiba BRT system, and there are new plans for its expansion, due to its success. Similar systems have been constructed in other Colombian cities. Currently, the system has completed its third construction phase out of a total of eight.

A clear definition of TransMilenio is given in [7]: "Trans-Milenio is defined as an urban mass transportation system that privately operates high capacity articulated buses that circulate through segregated busways, which are integrated into a system of feeder services that cover circular peripheral services with medium capacity buses. The system has stations with platform level boarding and automatic doors synchronized with the buses, where passengers take or get off the buses and the service is limited for those who have bus tickets. A satellite control system permanently supervises the buses, and the one-payment fee allows the passenger to board both busway and feeder services".

A. Infrastructure

TransMilenio's infrastructure is composed of three fundamental elements without which operation would be impossible: busways, stations and buses.

TransMilenio buses circulate on exclusive roads called *busways*. Currently the system has 11 busways with a total length of 104.6km ([8]). There are two types of busways: one lane and two lanes busways. The one-lane busways have passing lanes at stations so that buses can pass each other, thereby providing for express routes. The busways are located on the city's main roads and are physically separated from the mixed traffic lanes. The busways are in the central lanes of the roads [8], [7]. Internal rules of circulation and operation control can be imposed to improve system performance because the lanes are used exclusively by buses.

As a complement for the buses that circulate on busways, there are lower capacity buses that circulate on the other roads of the city. These routes are called *feeder routes* and have predefined stop points ([9]).

TransMilenio has a total of 143 *stations*. These stations form the area where the users can move and board bus routes (the "paid area"). The station platforms are at the same height as the bus doors, and that makes it easy to board the buses ([8], [7]). There are three types of stations: portal, intermediate and standard. The *portal stations* are the main stations of the system and are located at the ends of each busway. They are the starting and final destination points for the buses. Furthermore, they have access to feeder routes, which depart from and arrive at these stations like the busway routes. In this way, the transfer between routes becomes easier. The *intermediate stations* are similar to the portal stations (passengers can transfer between busway and feeder routes) but are located at intermediate points of the busways and not at the ends ([8]). The *standard stations* are smaller than the other two and only allow access for busway routes. They are located along the busways with an average separation of 500 meters. Their size varies and they can serve 1, 2 or 3 buses simultaneously in each direction.

TransMilenio has three types of *busway services*: normal, express and super-express. The *normal services* are routes that stop in all stations along the way. They have a higher frequency than the other types of services. The *express services* stop only at some stations (from 40% and 60%), and have a higher average speed than normal services. The stop plan of these services has been designed according to the levels of demand of the stations along the busway ([9]). The *super-express services* are very similar to the express services. The only difference is that they stop at fewer stations (about 20%) of the stations along the busway. These services are better for users that have long journeys because they stop at few stations ([9]). Currently, TransMilenio has 1392 busway buses (articulated or bi-articulated) and 574 feeder buses.

B. Speed

System buses operate at average speeds of 19 and 32 km/h for normal and express services, respectively ([9]).

TransMilenio increased the average speed of the city's corridors. For instance, the *Caracas* corridor had speeds of 12 km/h and *Calle 80* of 18 km/h due to the oversupply of private bus operators that there was before TransMilenio and which generated traffic jams.

TransMilenio system has an average speed of 26 km/h ([9], [1], [3]). This means that the average speed of public transport increased by 15 km/h after TransMilenio was implemented.

Nevertheless, the speeds are not the same in all busways for different reasons, such as the number of traffic lights along the busways, the number of lanes and even the type of material the street is made of [10]. For example, the *Eje Ambiental* busway, is a cobbled road in the historic center of the city. Table II shows the average bus speeds on the major busways of TransMilenio.

Busway	Speed (km/h)
Eje Ambiental (EW)	9.07
Eje Ambiental (WE)	10.51
Caracas (SN)	22.05
Caracas (NS)	22.61
Caracas Sur (NS)	24.82
Suba (SN)	24.95
Suba (NS)	25.08
Calle 80 (WE)	26.59
NQS Sur (NS)	27.32
Caracas Sur (SN)	28.17
Américas (EW)	28.24
Américas (WE)	28.37
Calle 80 (EW)	29.27
Autonorte (NS)	31.21
NQS Central (NS)	32.80
Autonorte (SN)	33.12
NQS Sur (EW)	33.18
NQS Central (SN)	36.87

TABLE II.   AVERAGE BUSWAY SPEEDS IN TRANSMILENIO

C. Capacity

TransMilenio has a maximum load capacity of 45000 passengers per hour per direction (pphpd), the highest-capacity BRT system in the world and even outperforming many heavy rail or metro systems ([1], [10]), see Table III.

Line	Type	Ridership (passengers/hour/direction)
Hong Kong Subway	Metro	80000
São Paulo Line 1	Metro	60000
Mexico City Line B	Metro	39300
Santiago de Chile La Moneda	Metro	36000
London Victoria Line	Metro	25000
Madrid Metro Line 6	Metro	21000
Buenos Aires Line D	Metro	20000
Bogotá TransMilenio	BRT	45000
Sã Paulo 9 de julho	BRT	34910
Porto Alegre Assis Brasil	BRT	28000
Curitiba Eixo Sul	BRT	10640
Manila MRT-3	Elevated rail	26000
Bangkok SkyTrain	Elevated rail	22000
Kuala Lumpur Monorail	Monorail	3000
Tunis	LRT	13400

TABLE III.   MAXIMUM CAPACITY OF MASS TRANSPORTATION SYSTEMS AROUND THE WORLD

The *transit capacity and quality of service manual* ([11]) defines the capacity of any route or public transport corridor as "the maximum number of people that can be carried past a given location during a given time period under specified operating conditions without unreasonable delay, hazard, or restriction, and with reasonable certainty". This capacity is measured in number of passengers per hour.

A system's capacity is limited by the component with least capacity (i.e., the bottle neck). The three key components of the BRT systems are buses, whose capacity is measured in number of passengers; stations, whose capacity is measured in number of passengers and buses; and busways, whose capacity is measured in number of buses. Whichever of these three components has the least capacity will become the factor that controls the system corridor. Several authors agree that BRT systems capacity is most often limited by the stations ([9], [10], [1]).

As mentioned before, TransMilenio has two types of buses that operate on the busways, articulated buses and bi-articulated buses. Table IV shows the number of passengers that each type of vehicle can carry in a one-lane busway and with one boarding platform stations, the average time that a vehicle occupies a given boarding platform (dwell time) and the average boarding time. TransMilenio increases the system capacity by using multiple boarding platforms in each station [1].

Type of vehicle	Maximum vehicle capacity (passengers)	Average dwell time (seconds)	Average bording & alighting time (seconds)	Corridor capacity (pphpd)	Vehicle capacity (vehicles/hour)
Articulated	160	13	0.3	9779	61
Bi-articulated	240	14	0.3	12169	51

TABLE IV.   VEHICLES AND PLATFORM CAPACITIES

Articulated buses carried an average of 1596 passengers in 2006, which is five times the average number of passengers carried by traditional buses. Furthermore, the number of kilometers a bus travels has increased due to the expansion of the busways, the extension of operating hours and the increase of express services. Each bus travelled 370 km daily in 2006 ([9]).

TransMilenio was the first BRT system to include multiple boarding platforms inside each station. In this way, it reached levels of capacity that only heavy rail systems had ([1]). Some TransMilenio stations may have up to five different platforms, each used for a different route.

There are reasons for including multiple platforms in a station [1]. The first one is to offer different types of services, such as normal and express, which can be allocated to different platforms. The second, and most important, is to reduce the saturation levels at stations, which helps to improve the service.

Besides, it is possible to distribute the different routes along each platform in such a way that each route stops only at one platform. It is then easier for users to find routes, because the user will associate each bus route with a platform.

In theory, one station with five platforms may have five times the capacity of a station with only one platform ([1]). To make this possible, the platform saturation level should be between 40% and 60%.

A TransMilenio capacity study was conducted in 2007 ([10]) and revealed which capacity values could be achieved according to the number of boarding platforms at each station, see Table V. Note that it is assumed that each platform has space to keep a vehicle in line (storage space).

Type of station	Recommended saturation (%)	Capacity (vehicles/hour)
Station with one boarding platform and no storage space	40	48
Station with one boarding platform with storage space	60	72
Station with two boarding platforms and no storage space	40 and 40	96=48+48
Station with two boarding platforms, one with storage space and the other one with no storage space	40 and 60	120=48+72
Station with two boarding platforms with storage space	60 and 60	144=72+72
Station with three boarding platforms where just one of them has storage space	40, 60 and 60	192=48+72+72
Station with three boarding platforms with storage space in each	60, 60 and 60	216=72+72+72
Station with four boarding platforms with storage space in each	60, 60, 60 and 60	288=72+72+72+72

TABLE V.   STATIONS CAPACITY ACCORDING TO THE NUMBER OF PLATFORMS ([10])

In BRT systems, the busway capacity is much higher than the station capacity. Bogotá's City Council *transit and transport administration manual* ([12]) states that the saturation flow of the busways is reached when there are between 692 and 750 articulated buses per lane. The interval is between 470 and 730 for right turns and between 465 and 735 for left turns. Note that this capacity does not contemplate elements such as intersections or traffic lights. It is clear that the busways capacity is much greater than the stations capacity.

The basic capacity of each busway is equal to the least capacity station along the busway ([10]). It does not account for questions that may increase system performance, such as express routes. The values for the different busways are (*Calle 26* and *Carrera décima* busways are not considered since they were opened after the date of the study) ([10]):

- *Caracas Centro* Busway: 192 buses/hour.
- *Autopista Norte* Busway: 144 buses/hour.
- *Avenida Suba* Busway: 144 buses/hour.
- *Calle 80* Busway: 48 buses/hour.
- *NQS* Busway: 72 buses/hour.
- *Américas/Calle 13* Busway: 144 buses/hour.
- *Eje Ambiental* Busway: 72 buses/hour.
- *Caracas Sur* Busway: 96 buses/hour.
- *Caracas Sur ramal Tunal* Busway: 72 buses/hour.

IV.   EXISTING STUDIES

There are very few proposals in the literature that focus on optimizing the BRT system operation. This can be explained because BRT systems are relatively recent (until the year 2000 there were only 19 BRT systems in the world ([4])). Another possible reason could be that many of the currently operating BRT systems are far from reaching their maximum capacity. For instance, none of the BRT systems operating in USA has reached maximum capacity and all of them have plenty of space for expanding their operation capacity ([6]).

Nevertheless, there are very interesting proposals that can be used as a starting point to propose a model for optimizing the operation of TransMilenio. To the best of our knowledge, there are no automatic route design proposals. The closest to this is the model proposed in [2]) that evaluates and selects the best several routes.

In the following we review of different proposals in the literature to manage and optimize the operation of different BRT systems and, specifically, for TransMilenio.

A. Proposals for optimizing of BRT systems

Most of the proposals in the literature for optimizing BRT systems are based on bus scheduling and are focused on varying the times between each bus departure (i.e., the headway) of the different bus routes.

In [13] a model for optimizing BRT systems is proposed on the basis of two elements, the headway, which is assumed to be uniform, and the order in which the bus routes depart.

The optimization model is characterized by a set of predefined bus routes (normal, zone and express routes). A random number is generated and assigned to the headway. Then, the algorithm finds an optimal solution to the order in which the routes have to depart that minimizes a cost function. For instance, the algorithm may determine that for a headway of 5 minutes the best departure order is [normal, express, express, normal, zone]. This solution means that a normal bus route should depart at minute 0, an express route at minute 5, a normal route at minute 10 and an express route at minute 15 and a zone route at minute 20. This solution may be better than for example [express, express, zone, express].

The cost function accounts for the passengers waiting at the stations, the waiting time inside the buses and the operating costs. It is very complete and includes several variables, such as the number of boarding/alighting passengers by station, the stops of each route, the monetary value of the waiting time costs and the vehicle operating costs, among others.

The model chooses a headway for the given routes and shuffles the order of departure. A genetic algorithm is used to reach an optimal solution. The article presents a novel codification that is a vital element for the model and includes a combination of the headway and route design variables.

In [14] a very complete model is proposed, with good granularity and with greatly detailed costs. An application to Line 2 of the BRT in Beijing is used to illustrate the model. The total cost of a solution accounts again for passenger waiting time at the station, passenger waiting time inside the bus, and the vehicle operating costs. Passenger walking time from home/office to the station is excluded, because bus scheduling has no influence on that time.

The model considers variables such as the bus departure frequency, the distance between stations, average speed between stations, the rates and boarding times, acceleration and deceleration times, the number of traffic lights between stations, the traffic lights cycle times and others. Furthermore, it assumes that the waiting time is equal to half of the frequency time or headway.

Fixed costs are removed from the analysis because they are unaffected by bus scheduling. The variable costs are composed of operating cost per kilometer, operating hours, vehicle depreciation, etc.

The decision variables in the optimization model are the route headway and binary variables that represent whether or not stations are skipped. The model seeks to minimize the total costs and is subject to capacity constraints, vehicle availability and headway limitations. To accomplish this of an all-stop route and an express route is combined and their headways are calibrated to minimize the total costs.

The algorithm complexity increases exponentially along with the number of stations, and this is the main reason why the authors use a genetic algorithm (it would be too costly to use a deterministic algorithm). Another reason its that genetic algorithms are able to naturally represent binary variables.

[2], [15], [16] introduce an optimization model for the minimization of waiting time, travelling time and operating costs for an express bus service, given the travel demand. A mathematical model is built to minimize costs given a set of stations, the distance between stations, the passenger origin-destination matrix and a set of *a priori* attractive set of routes. For each suggested route the model outputs the frequency of the services and the size of the buses to use.

For the construction of this model the travel demand is assumed to be fixed and known, represented by a origin-destination matrix for the analyzed stations, which must be satisfied. It is also assumed that passengers arrive at an average fixed rate, passengers choose the route to their destination that minimizes travel time and there is no limit on the available vehicle fleet.

The operating costs are computed on the basis of the cycle cost of a full bus route, the frequency of each route and the operating set of routes. The passengers costs are given by the waiting time at stations, travel time and transfer time.

There are also various proposals to improve the operation of BRT systems through the prioritization of transit signals. [17] describes the mathematical relationship between the departure frequency of a route, the cycle length of the transit signals and the number of different signal states when the buses arrive at an intersection. It proposes various strategies for prioritizing signals that decrease the headway time deviation, i.e., decrease the punctuality deviation of the buses, without having a significant impact in the delay of the mixed traffic.

Other proposals that study the priority control of signals in BRT systems can be found in [18], [19], [20], [21].

B. Proposals for the optimization of TransMilenio

Since the construction of TransMilenio in the year 2000, various proposals have been made to optimize its operation. Most of these proposal focus on the reduction of systems costs, to strike a balance between passengers waiting time and operating costs.

For instance, Petri networks are used to model TransMilenio in [22]. The proposed model is classified as a macroscopic deterministic simulation model, due to its detail level, process and operation representation. The model uses a multiagent approximation to model three important system components: the passenger behavior (how many passengers take the bus per hour), the busway dynamics, and the interaction between the passengers and the buses. Since Petri networks are unable to deal with time, trigger times are added to the nodes to represent temporal relationships.

Three busways are modeled, *Avenida Caracas*, *Autonorte* and *Avenida de las Américas*. The model includes the seven most important stations out of a total of 45 on these busways. Moreover, three routes (a normal route, an express route and a super-express route) that stop at the same stations on their back and forth trips were chosen.

Two Petri networks are designed. One models the whole system and randomly assigning buses to routes, and the other separates the routes from the buses. Random models have the advantage of being able to simulate the interaction between routes. The random model outperforms non-random models to satisfy the demand with the minimum number of resources. Finally, the random model works as an integrated system and it is capable of solving perturbation by itself. The result of this simulation shows that there is a point at which adding more buses to the system does not improve the performance.

In [23] a genetic algorithm is used to find the best frequency for pre-established bus routes that minimizes passenger waiting time. The frequency is determined by the assignment of buses to each route. The model tries to minimize the time the passengers spend on the system, which is composed of the travel time plus the waiting time at the stations.

The genetic algorithm chromosome size is equal to the number of routes and the population is initialized randomly with the constraint that each route has at least one bus assigned. A random matrix is also created along with the initial population, this matrix contains all origin-destination trips. This algorithm assumes the user is "smart" and will always choose the best route to go to his/her destination.

The arrival of buses at the station and the passengers waiting time are modeled by a Poisson process and a distributed Erlang event, respectively. It accounts for the scenario where buses are full and passenger cannot board. In these cases the passengers have to wait to the next bus.

A graph with the routes was designed to measure the time, where each node represents a station and the arcs represent the connections between them. The arc costs are the travel time between the stations that the arcs connect. Additionally, arcs with the possible express routes are included. Dijkstra's algorithm is then used to compute the shortest routes, and it is executed before running the algorithm.

In [24] a model to evaluate TransMilenio routes is built based on the data provided by a origin-destination matrix. The trip probabilities between stations and passenger arrival rate to each station are computed from this matrix, assuming that the users know which is the best route to reach their destination. The model is implemented in a commercial simulation software package.

The model includes a set of constraints regarding user behavior when choosing their route to reach to their destination. The input data for the evaluation algorithm are the origin-destination pairs, the stops of each service and the quantity of passengers associated with each pair. TransMilenio data is also required, such as existing routes, their frequency, vehicle capacity, speed, the distance between stations and other network characteristics.

The model is composed of three modules. The network module stores information about the physical infrastructure, such as stations, the distance between them, busways, and others; the stations module is in charge of the boarding and alightment at each station and the arrivals module assigns passenger origin and destination.

The time that the passenger spends inside the station is given by the travel time, the bus stopping time and passenger waiting time at the station. The bus arrival times are assumed to be uniform. Therefore, the waiting time of each passenger is equal to half of the route's headway.

C. Analysis of proposed models

Most of proposed models are far from being able to represent what goes on in the real world because they are not detailed enough to represent what happens within a BRT system. We are going to describe the advantages and drawbacks of each model.

The model presented in [23] does not account for vehicle operating costs. This fact is clearly reflected in the results, where the best solution is to increase bus departure frequencies and use the entire bus fleet. The models do not adequately represent constraints concerning capacities within the system, which is modeled for the whole corridor but not for each station individually. This overlooks the fact that there are some stations that have more demand than others. The model does not account for deceleration and acceleration times, passengers boarding and alighting times or dwell times at signalized intersections. An advantage of this proposal is that it builds a graph that pre-calculates the travel times between each pair of stations for each route. This is helpful to find the optimal route between two stations.

The model presented in [22] has several voids, such as the fact that it does not account for passenger waiting time or vehicle operating costs. Neither does it account for passenger congestion within the stations, vehicle congestion at stations, discriminated speeds between each pair of stations, or assign distribution times to passenger and vehicle arrivals at stations. On the other hand, it has several advantages, such as considering that when buses are full passengers must wait for the next bus. The model is a user-friendly graphic tool that can model a system in which equations are not known.

The model presented in [24] refers to some important constraints but the model does not include any. Other constraints included as assumptions are not necessarily realistic. For instance, it is assumed that if a passenger is going to make a trip that is 5 or less stations of long, he/she will only take normal (all-stops) services. It does not consider decelerating and accelerating times, passenger boarding and alighting times or dwell times at signalized intersections either.

The model presented in [13] assumes the same speed between every pair of stations, which is not realistic. It does not consider the bus passenger capacity, vehicle capacity at the stations and passenger capacity at the stations either. Waiting times at signaled intersections are not considered either. A major drawback of this model is that it uses the number of passengers that board and alight from buses at each station rather than an origin-destination matrix as input. This demand data is not detailed enough to identify passenger behavior. The model accounts for passenger waiting times and vehicle operating costs. The introduction of an innovative variable-size codification and the use of binary variables to indicate whether or not a bus stops at a station.

The model presented in [14] is one of the most complete. In fact, it incorporates most cost variables. It is the only model that includes passenger boarding and alighting times and the stop times at the signaled intersections. Nevertheless, it has some drawbacks. For example, it only considers one express route, i.e., scenarios with several express routes cannot be evaluated. The model is aimed at reaching the best departure frequency for a normal and an express route that operate along the same busway. Its parameters are the origin-destination matrix, the stations in which the express routes stops and the bus fleet size. Note, finally that the model does not consider vehicle congestion or passenger congestion at stations.

The model presented in [2] is also very complete and perhaps the best at representing the costs of a real BRT system. This is accomplished thanks to the inclusion of several express routes on one busway and because it is good at differentiating travel time and operating costs. But it is not free of drawbacks. For instance, it does not include acceleration and deceleration times, boarding and alighting times or stop times at signaled intersections. The proposed model searches the departure frequencies that optimize BRT operation according to a defined cost function.

In summary, none of the reviewed proposals considers vehicle congestion at stations or passenger congestion at stations. This is worrying, because, as stated in [9], [10], [1] the capacity bottleneck of a BRT system is the vehicle capacity at the stations. There are not many proposals that account for this point because hardly any BRT systems have reached maximum capacity, which could be the reason why the proposals have focused mainly on the minimization of passenger waiting times an operating costs, and not on the increase of system capacity.

We found that none of the proposals offer automated route design. [14] and [2], which offer validation models for routes that can be given to the model as a parameter, come the closest. We also found that none of the proposals take a multi-objective approach to the problem.

V. PROPOSED OPTIMIZATION MODEL FOR TRANSMILENIO

In the previous section we reviewed the proposals in the literature for optimizing BRT system operation, with an special emphasis on the TransMilenio system. In this section, we provide a mathematical model for the optimization of Trans-Milenio that adapts elements of the above proposals, mainly [14] and [2], and incorporates novel elements accounting for the features of that system.

The problem is to find departure frequencies for the established routes that minimize the time passengers spend inside the system and operating costs. This set of frequencies must satisfy the constraints associated with the TransMilenio operation. The problem is analysed only during the rush hour time window.

A. Available information

The *set of stations* determines the size of the BRT system. The number of stations is directly related to the complexity of the problem to be solved. Information about the system stations, the busways to which they belong, and each station's neighboring stations must be considered. The set of TransMilenio stations is denoted by $E = \{e_1, \ldots, e_{143}\}$, where e_i refers to the i-th station, $i = 1, \ldots, 143$.

The *routes* are paths between two stations (usually main stations) that buses must take and are composed of the set of station at which buses must stop. The set of TransMilenio routes is denoted by $R = \{r_1, \ldots, r_{90}\}$, where r_j is the j-th route, $j = 1, \ldots, 90$.

The *station vehicle capacity* is very important, even more so in cases where nearing full capacity the BRT system is, like the TransMilenio system. When the system is nearly at maximum capacity the problem is to find feasible solutions that can meet the trip demands. The station vehicle capacity e_i is denoted by $k^s_{e_i}, i = 1, \ldots, 143$.

The information required about the *buses* is their capacity and the quantity of buses in operation. We assume that the buses operating along each route have the same capacity. This information is important in order to impose capacity constraints within the buses and to prevent to operate with more buses than the available. The bus capacity is denoted by $k^b_{r_j}, j = 1, \ldots, 90$ which is the passenger capacity of the vehicles that operate the r_j-th route.

The *distance between stations* is used to compute the travel times between each pair of stations. $d_{e_i e_j}$ denotes the distance between stations e_i and e_j, $i, j = 1, \ldots, 90$.

Speeds between each pair of stations are very important because not all busways have the same characteristics and therefore the speed is not always the same. Some busways have signaled intersections, whereas others are built over highways where they can travel at faster speeds. The speed between the stations e_i and e_j is denoted by $s_{e_i e_j}$.

Acceleration and deceleration times along with boarding and alighting times are used to compute the total time of a stop at a station. These values are constant and independent of passenger demand level in the system. Based on the model proposed in [14], we assume that the times are the same for all stations. The acceleration and deceleration times are denoted by p.

The *boarding and alighting times* are used to determine the total stop time of a bus at a station. The stop time increases with the amount of people that board or alight the bus. Based on the model proposed in [14], we can calculate passenger boarding time at a station as $\alpha^{r_j}_{e_i} \times \tau^\alpha$, where $\alpha^{r_j}_{e_i}$ is the passenger boarding rate for route r_j at station e_i and τ^α is the passenger boarding time. In the same way, the alighting time is denoted by $\beta^{r_j}_{e_i} \times \tau^\beta$, where β is used for alightings.

Costs are usually divided into passenger waiting time costs and the BRT system operating costs. The model that we propose accounts for three types of costs: waiting time at stations, waiting time on buses and vehicle operating costs. Fixed costs, such as station cleaning, electricity, administrative wages, rents, and others, are not considered because they are independent of the BRT system operation ([14]). The unit cost per kilometer, the unit cost for waiting time at the station and the unit cost for waiting time inside the buses, are denoted by μ_O, μ_S and μ_B, respectively. These values are used in the cost function to evaluate the quality of the sets of routes.

The *origin-destination matrix* contains information about passenger demand, i.e., the amount of users traveling from station e_i to station e_j. We use an origin-destination matrix with rush hour data, because this is the time window when the system is closer to maximum capacity. The number of passengers that travel from station e_i to station e_j is denoted by $q_{e_i e_j}$.

Operating hours is the time during which the BRT system is operating, denoted by T.

B. Decision variables

The decision variables for the proposed model are the the frequencies associated with each route. The set of frequencies is denoted by $F = \{f_1, \ldots, f_{90}\}$, where f_{r_k} is the frequency for the buses of the k-th route. The frequencies identify how often the buses of a given route depart. The headways can be computed from the frequencies and vice-versa.

C. Cost function

Multiple authors (see [14], [2], [13]) agree that the cost function, C, is composed of the sum of three elements: vehicles operating costs, C_O; passenger waiting time at station costs, C_S, and passenger travelling time costs, C_B. These last two costs can be grouped as the passenger total trip costs ([14]).

Then, the function to be optimized (minimized) is:

$$minC = C_O + C_S + C_B,$$

The operating costs can be calculated by:

$$C_O = \mu_O \times \sum_{r_k \in R} T \times f_{r_k} \times D_{r_k},$$

where μ_O is the unit cost per kilometre for a BRT vehicle, R is the set of all routes in the system, T is the BRT system operating hours, f_{r_k} is the frequency of route r_k and D_{r_k} is the length of the path covered by route r_k. D_{r_k} can be computed from the distances ($d_{e_i e_j}$) between the consecutive stations included in the k-th route.

The waiting time at station costs can be computed as follows:

$$C_S = \mu_S \times \sum_{e_i, e_j \in E} q_{e_i e_j} \frac{\epsilon}{\sum_{r_k \in R} f_{r_k} \times x^{r_k}_{e_i e_j}},$$

where μ_S is the waiting time unit cost, $q_{e_i e_j}$ is the passenger trip demand for the (e_i, e_j) origin-destination pair, f_{r_k} is the frequency of route r_k, $x^{r_k}_{e_i e_j}$ is a binary variable that indicates whether a route r_k is a good option for travelling from the station e_i to the station e_j (its value is 1 if the route is attractive and 0 otherwise), and ϵ is the bus arrival distribution at the stations, which are assumed Poisson distributions.

The travel time costs can be computed by:

$$C_B = \mu_B \times \sum_{e_i, e_j \in E} q_{e_i e_j} \times \frac{\sum_{r_k \in R} t^{r_k}_{e_i e_j} \times f_{r_k} \times x^{r_k}_{e_i e_j}}{\sum_{r_k \in R} f_{r_k} \times x^{r_k}_{e_i e_j}},$$

where μ_B is the travel time unit cost, f_{r_k} is the frequency of route r_k, and $t^{r_k}_{e_i e_j}$ is the travel time in route r_k for the (e_i, e_j) origin-destination pair, with

$$t^{r_k}_{e_i e_j} = t^{1 r_k}_{e_i e_j} + t^{2 r_k}_{e_i e_j},$$

where $t^{1 r_k}_{e_i e_j}$ and $t^{2 r_k}_{e_i e_j}$ are the travel time and delay time from station e_i to station e_j,

$$t^{1 r_k}_{e_i e_j} = \frac{d_{e_i e_j}}{s_{e_i e_j}},$$

$$t^{2 r_k}_{e_i e_j} = \left(\frac{\sum_{e_i \in P_{r_k}}^{e_j} \alpha^{r_k}_{e_i} \times \tau^{\alpha}}{f_{r_k}} + \frac{\sum_{e_i \in P_{r_k}}^{e_j} \beta^{r_k}_{e_i} \times \tau^{\beta}}{f_{r_k}} \right) + N_{e_i e_j} \times d,$$

and

- $d_{e_i e_j}$ and $s_{e_i e_j}$ are the distance and average speed between station e_i to station e_j, respectively,
- $\alpha^{r_k}_{e_i}$ and $\beta^{r_k}_{e_i}$ are the boarding and alighting rates at station e_i for route r_k, respectively,
- τ^{α} and τ^{β} are the boarding and alighting times per passenger,
- $N_{e_i e_j}$ is the number of stations between station e_i and station e_j,
- P_{r_k} is the set of stations at which route r_k stops,
- p is the acceleration and deceleration delay at station.

D. Constraints

The *bus passenger capacity* constraint ensures that the frequency of bus departure is high enough to prevent overcrowding inside the buses. If this constraint is not applied, buses may be full when the arrive at stations, passengers will have to wait for the next bus.

$$k^b_{r_k} \times f_{r_k} \geq \sum_{e_i \in P_{r_k}}^{a} \sum_{e_j \in P_{r_k}}^{N} q_{e_i e_j} \times \frac{f_{r_k} \times x^{r_k}_{e_i e_j}}{\sum_{r_m \in R} f_{r_m} \times x^{r_m}_{e_i e_j}},$$

$\forall r_k \in R, \quad \forall a \in P_{r_k}$, where

- $k^b_{r_k}$ is the passenger capacity of the buses circulating along route r_k,
- f_{r_k} is the frequency of route r_k,
- P_{r_k} is the set of stations at which route r_k stops,
- $q_{e_i e_j}$ is the passenger trip demand for the (e_i, e_j) origin-destination pair,

- $x^{r_j}_{e_i e_j}$ indicate whether a route r_j is a good option for travelling from the station e_i to the station e_j. Its value is 1 if the route is attractive and 0 otherwise.

The *bus fleet size* constraint prevents the set of routes from operating with more buses than are available in the system. This assures that the system is working with the available resources:

$$\frac{T}{N^{r_j}} \leq \frac{1}{f_{r_j}}, \quad \forall r_j \in R,$$

where T is the BRT system operating hours and N^{r_j} is the number of vehicles that can operate along the route r_j.

The *choice of best route* constraint helps to model passenger behavior when choosing a route to travel to their destination. It models the possibility of passengers often being able to take more than one route to reach their destination in the same time.

$$x^{r_k}_{e_i e_j} = 1$$
$$\Updownarrow$$
$$\mu_B \times t^{r_k}_{e_i e_j} \leq \frac{\mu_S + \mu_B \times \sum_{r_m \neq r_k} t^{r_m}_{e_i e_j} \times f_{r_m} \times x^{r_m}_{e_i e_j}}{\sum_{r_m \neq r_k} f_{r_m} \times x^{r_m}_{e_i e_j}},$$

$\forall r_k \in R, \forall e_i, e_j \in E.$

The *station vehicle capacity* constraint is very important especially in systems that have great passenger demands and are nearing maximum capacity level. The importance of this constraint is that the vehicle station capacity is the bottleneck of the BRT systems ([9], [1]), like TransMilenio.

$$k^s_{e_i} \geq \sum_{r_j \in R} v^{r_j}_{e_i} \times f_{r_j}, \quad \forall e_i \in E,$$

where $k^s_{e_i}$ is the vehicle capacity of station e_i, and $v^{r_j}_{e_i}$ are binary variables that point out whether station e_i is visited on the route r_j.

VI. CONCLUSIONS AND FUTURE RESEARCH

TransMilenio is Bogotá's most important mass transportation system and one of the biggest BRT systems in the world. There are very few proposals in the literature that focus on optimizing BRT system operation, mainly because BRT systems are relatively a recent form of transport and many of the currently operating BRT systems are nowhere near full capacity.

Most of proposals, and specifically for TransMilenio, are based on bus scheduling and focus on varying the times between each bus departure (i.e., the frequencies) of the different bus routes to minimize costs.

In these proposals, the set of routes are part of the available information, along with the stations at which the buses stop, and they remain constant during the execution of the model.

In the mathematical modeling introduced in this paper we also analyze the frequencies of the routes to minimize costs.

Note that an automated design of routes that minimize the cost function is an open research line. Rather than designing new routes, the aim would be to optimize existing routes by modifying at most 30% of the stations on the original route. This 30% was fixed by TransMilenio experts at meetings. The reasons for just modifying rather than redesigning routes is that the social impact of modifying the routes is not too high, whereas, the search space is greatly reduced and, therefore, better solutions can be found in less time. An important drawback is that it may not be possible to find a global optimum, because the best routes may have less than 70% of the stations in common with the original routes. In this case it is more important to reduce the social impact on passengers that comes with the modification of the routes.

We are now working with TransMilenio experts on extending and solving the proposed optimization problem. We have selected evolutionary algorithms to solve the problem since they have previously proven to be efficient tools. Additionally, the research team is experience in solving other complex optimization problems using this metaheuristic.

The model we propose is a single objective optimization model since only costs are minimized. However, other objectives could be simultaneously considered, leading to a multi-objective optimization model. Evolutionary algorithms would be then used to identify Pareto optimal solutions, and the expert's preferences could be incorporated into the search process to reach a compromise (satisficing) solution.

Another future research line that we propose is the possibility of adding transfer times to the model. Transfer time is the time it takes to a passenger to switch from one route to another, usually because the first bus that a passenger takes does not stop at the station for which he or she is heading. These times are normally penalized because transfers are an inconvenience for passengers.

Finally, another open research line, and a key aspect for correctly modeling BRT systems is user's behavior. It is important to correctly model which routes users given an origin/destination pair will choose. They are likely to choose the fastest route, but this is not always the case, because users may not know which the fastest route is or because the frequency of the fastest route is low and they opt for an alternative route. This is one of the least explored issues in the BRT systems literature.

ACKNOWLEDGEMENT

The paper was supported by Madrid Regional Government project S-2009/ESP-1685 and the Spanish Ministry of Science and Innovation project MYTM2011-28983-C03-03. The authors would also like to acknowledge the Ministry of Information and Communication Technologies of Colombia and its Talento Digital scholarship program for funding the studies of Francisco J. Peña, and Alirio García for the useful information he provided about TransMilenio system.

REFERENCES

[1] L. Wright and W. Hook, *Bus rapid transit planning guide*. Institute for Transportation & Development Policy New York, 2007.

[2] C. Leiva, J. C. Muñoz, R. Giesen, and H. Larrain, "Design of limited-stop services for an urban bus corridor with capacity constraints," *Transportation Research Part B: Methodological*, vol. 44, no. 10, pp. 1186–1201, 2010. [Online]. Available: http://www.sciencedirect.com/science/article/pii/S0191261510000044

[3] Observatorio de Movilidad, "Comportamiento de los indicadores de movilidad de la ciudad a diciembre de 2010," Observatorio de Movilidad, Tech. Rep. 6, 9 2011.

[4] (2013, 2) Panoroma of brt and bus corridors in the world. BRTdata.org. [Online]. Available: http://brtdata.org/

[5] H. Levinson, S. Zimmerman, J. Clinger, S. Rutherford, R. L. Smith, J. Cracknell, and R. Soberman, "Bus rapid transit, volume 1: Case studies in bus rapid transit," Transportation Research Board, Tech. Rep., 2003.

[6] D. Hinebaugh, "Characteristics of bus rapid transit for decision-making," Tech. Rep., 2009.

[7] I. Chaparro, *Evaluación del impacto socioeconómico del transporte urbano en la ciudad de Bogotá: el caso del sistema de transporte masivo, TransMilenio*, ser. Serie Recursos Naturales e Infraestructura. United Nations Publications, 2002. [Online]. Available: http://www.eclac.org/publicaciones/xml/3/11423/LCL1786-P-E.pdf

[8] S. L. Ángel, C. García, O. Santiago, J. Concha, and M. R. Caldas, "Plan Marco 2010 Sistema TransMilenio," Dirección de Planeación de Transporte, Tech. Rep., 2010.

[9] A. Cain, G. Darido, M. R. Baltes, P. Rodriguez, and J. C. Barrios, "Applicability of Bogotá's TransMilenio BRT System to the United States," National Bus Rapid Transit Institute (NBRTI); Center for Urban Transportation Research (CUTR); University of South Florida, Tech. Rep., 2006.

[10] Steer Davies Gleave, "Estudio de determinación de la capacidad del sistema TransMilenio. Estudio realizado para TransMilenio SA, Bogotá," Steer Davies Gleave, Tech. Rep., 2007.

[11] Kittelson & Associates and United States. Federal Transit Administration and Transit Cooperative Research Program and Transit Development Corporation, *Transit Capacity and Quality of Service Manual*. Transportation Research Board National Research, 2003, vol. 100.

[12] A. F. G. Valderrama, S. H. Pérez, M. C. Benítez, J. C. Grisales, M. A. Hinojosa, D. E. Duenas, J. C. T. Gomez, L. A. Guzmán, N. C. Acosta, M. V. Ropero, and P. J. G. H. Edna Rodríguez Aleman, *Manual De Planeación Y Diseño Para La Administración Del Tránsito Y Transporte*, E. C. de Ingeniería, Ed. Alcaldía Mayor de Bogotá, 2005.

[13] C. Sun, W. Zhou, and Y. Wang, "Scheduling combination and headway optimization of bus rapid transit," *Journal of Transportation Systems Engineering and Information Technology*, vol. 8, no. 5, pp. 61–67, 2008.

[14] X. Chen, B. Hellinga, C. Chang, and L. Fu, "Optimization of headways for bus rapid transit system with stop-skipping control," in *Transportation Research Board 91st Annual Meeting*, no. 12-1999, 2012.

[15] H. Larrain, R. Giesen, and J. C. Muñoz, "Choosing the right express services for bus corridor with capacity restrictions," *Transportation Research Record*, no. 2197, pp. 63–70, 2010.

[16] H. Larrain, R. Giesen, and J. C. Munoz, "The effect of od trip dispersion versus concentration in express service design," in *12th World Conference on Transport Research. Lisbon, Portugal, 11-15 July 2010*, 7 2010. [Online]. Available: http://intranet.imet.gr/Portals/0/UsefulDocuments/documents/03406.pdf

[17] W. Ma and X. Yang, "A passive transit signal priority approach for bus rapid transit system," in *Intelligent Transportation Systems Conference, 2007. ITSC 2007. IEEE*, 2007, pp. 413–418.

[18] K. Wang and F. Zhu, "A real-time brt signal priority approach through two-stage green extension," in *9th IEEE International Conference on Networking, Sensing and Control (ICNSC), 2012*, 2012, pp. 7–11.

[19] L. Yong, Y. Yun, and G. Hongli, "The design of the brt signal priority control at the intersection," in *International Conference on Intelligent Computation Technology and Automation (ICICTA), 2008*, vol. 1, 2008, pp. 507–511.

[20] S. Cao, L. Zhang, and F. Zhao, "An rfid signal capturing and controlling system for brt priority system," in *Proceedings of the 2009 First IEEE International Conference on Information Science and Engineering*, ser. ICISE '09. Washington, DC, USA: IEEE Computer Society, 2009, pp. 4725–4729. [Online]. Available: http://dx.doi.org/10.1109/ICISE.2009.293

[21] H. Xu, J. Sun, and M. Zheng, "Comparative analysis of unconditional and conditional priority for use at isolated signalized intersections,"

Journal of Transportation Engineering, vol. 136, no. 12, pp. 1092–1103, 2010.

[22] D. Lopez, A. Triana, and H. Chamorro, "Simulation model of public transportation system using multiagent approach by means of petri nets: Bogotá study case," in *Robotics Symposium, 2011 IEEE IX Latin American and IEEE Colombian Conference on Automatic Control and Industry Applications (LARC)*. IEEE, 2011, pp. 1–6.

[23] S. Duarte, D. Becerra, and L. Niño, "Un modelo de asignación de recursos a rutas en el sistema de transporte masivo TransMilenio," *Avances en Sistemas e Informática*, vol. 5, no. 1, pp. 163–172, 2008. [Online]. Available: http://www.revista.unal.edu.co/index.php/avances/article/download/9984/10516

[24] M. A. Valbuena and D. Hidalgo, "Propuesta metodológica para la evaluación de rutas del sistema TransMilenio," 2005.

Optimization in Smallholders' Agricultural Farm: A Case of Goat Rearing in Malaysia

[1]Mohd Sahar Sauian, [2]Norfarhanah Che Othman
Faculty of Computer and Mathematical Sciences
Universiti Teknologi MARA
40450 Shah Alam, Malaysia.
[1]mshahar@tmsk.uitm.edu.my, [2]farhanah.co@gmail.com

Abstract—**Operations research problems were normally based on the optimum allocation of resources. The application is extensively used in various fields of research and development, including management, transport, finance and agriculture. Some of the common approaches available to solve such problems are linear programming, goal programming, integer programming, network and scheduling. In Malaysia, rearing of goats by farmers who were classified as small and medium entrepreneurs, were faced with many problems, particularly with respect to higher costs and diseases. In retrospect, goat rearing is not as profitable as compared to cattle rearing. This paper highlights the use of optimization tool in balancing demand of goats in livestock rearing. A goal programming approach is utilized in order to plan and satisfy seasonal as well as normal unseasonal demand. The model could also take into consideration the difference in priorities of the demand for different types of goats during the festive and non-festive seasons. The results suggest the practicality of the model in determining the optimal combinations of various types of goats to be supplied in order to satisfy the targeted goals. Hence, applying this tool would give beneficial impact to the entrepreneurs as it could be used as a model for managing small agricultural farms.**

Keywords: optimization; resource allocation; linear programming; MCDM; goal programming; multiple goals.

I. INTRODUCTION

Optimal utilization of resources has been the fundamental aspect in the application of operations research. Various approaches of operations research techniques had been utilized in solving problems in manufacturing, distribution, logistics, scheduling and even in management. One of the approaches includes the linear programming approach which has been widely used in multifarious fields of research and development (RD). The main idea is to optimize the use of resources in order to gather maximum returns. In recent years, the use of MCDM (multiple-criteria decision making) approach has been growing both in theory and practice.

MCDM is seen as a better optimization tool as it can handle problems with multiple objectives and it can also take into consideration different measurements as well. As such, MCDM allows appropriate selection of resource allocation strategies depending on the different objectives and management `styles' of particular individuals or households [1]. An example related to the use of MCDM is a case of Norwegian meat production industry where mixed integer programming model was used to optimize livestock allocation. The model successfully integrates components of supply chain management, routing of supply vehicles and production scheduling at the abattoir, where production plans determine transportation as well [2]. Another aspect of MCDM is used in marketing and production planning using a goal programming approach [3].

II. LIVESTOCK SECTOR IN MALAYSIA

The agricultural sector is one of the twelve potential National Economic Areas (NKEAs) that have been listed to drive economic growth in the tenth Malaysia Plan. During the period (2011-2015), the sector is expected to contribute up to 7.6% of GDP [4]. The increasing demand was particularly prominent in the ruminant sub-sector (cattle and goat).

The Malaysian demand for goats were particularly due to mutton, aqiqah and qurban as well as dairy. From 2006-2010, goat rearing industry showed a modest growth of 6.76% per year [5]. However, Malaysia still depends on imported goat from oversea such as Australia, New Zealand, Indonesia, South Africa and Thailand in order to meet the demand.

The Ministry of Agriculture and Agro-Based Industry (MOA) gave special focus on strengthening the ruminant sector by formulating strategies to increase the supply of good quality of cattle and goats. This was done by implementating commercial cattle and goat husbandry projects through the government and the private sectors.The move done by the government is to increase the competitiveness of the agricultural sector by maximizing the income through optimum utilization of resources, especially for smallholders farms [6].

III. OBJECTIVE

The objective of the study is to apply the goal programming approach in optimizing the use of resources in a goat farm in Selangor, Malaysia.

IV. METHODOLOGY

The methodological approach in dealing with this problem is through the utilization of goal programming. It seems appropriate to use this method due to the existence of many goals as well as the different measures of the goal themselves.

Various examples of goal programming approaches were applied in manufacturing industries, finance, operations management as well as in agriculture. Rehman, Lara and Romero (1992), for instance, used goal programming in the optimization of livestock rationing [7]. Kwak, Schneiderjans (1993) used goal programming approach in marketing distribution [8]. On the other hand, Sauian (2006) used the approach in strategizing production in a plastic factory, while Vitoriano, Ortuno and Tirado (2009) utilized the approach in humanitarian aid distributions [9, 10].

The distinguishing feature of the goal programming approach as compared to ordinary linear programming is that, it tends to minimize the deviations from the given goals.

V. APPLICATION OF GOAL PROGRAMMING AT KTP FARM

The company KTP Farm, focuses on rearing cattle and goat commercially using feedlot system to meet the local demand. The cattle farm has less problem as it is always profitable to manage. However, for managing the goat farm it needs proper attention in the use of resources as quite often medical costs tend to fluctuate as goats are prone to contact diseases. The goat's farm is capable to hold up to 600 goats under 15 pans or barn at a time. However, they are around 5 pans reserved for the kid and lamb yeanlings. There are five types of goats named Boer, Jamnapari, Katjang, sheep and cross breed Boer-Jamnapari reared by KTP Farm.

Given the various information supplied by the management of KTP Farm, we can formulate the relevant optimization goal programming model.

A. Goal Programming Model at KTP Farm

The goal programming model for a non-festive month is given as follows:

Minimize $Z = P_1d_1^- + P_2d_2^- + P_3d_3^+ + P_4d_4^+$ (1)

Subject to:

$\sum c_jx_j + d_1^- - d_1^+ = \prod$ (j=1,..,5) (goal 1)   (2)

$\sum x_j + I_{t-1} - d_2^- - d_2^+ = H$ (j=1,..,5) (goal 2)   (3)

$\sum a_jx_j + d_3^- - d_3^+ = G$ (j=1,...,5) (goal 3)   (4)

$\sum b_jx_j + d_4^- - d_4^+ = \Omega$ (j=1,...,5) (goal 4)   (5)

$\sum e_jx_j \leq h$ (j=1,...,5)   (6)

$\sum k_jx_j \leq t$ (j=1,...,5)   (7)

$x_j \leq L_j$ (j=1,..,5)   (8)

$x_j \geq U_j$ (j=1,..,5)   (9)

and $x_j \geq 0$ (for j=1,...5), $d_i^+, d_i^- \geq 0$ (for i=1,..,4)   (10)

where:

P_1, P_2, P_3, P_4 : priorities to the respective goals,

x_j : number of goat j supply from the farm,

c_j : profit per unit of goat j,

a_j : feeding cost per unit of goat j,

b_j : medical cost per unit of goat j,

e_j : labor cost per unit of goat j,

k_j : overtime cost per unit of goat j,

I_{t-1} : stock of goat from previous month,

$\prod$: targeted profit,

H : targeted demand of goats,

G : targeted total feeding cost for goats,

Ω : targeted total medical cost for goats,

j : types of goat (1=Boer goat, 2= Jamnapari goat, 3=Kitjang goat, 4= sheep, 5=cross-breed Jamnapari-Boer goat),

h : maximum labor cost available per month,

t : maximum overtime cost per month,

L_j : lower bounds for total number of goats,

U_j : upper bounds for total number of goats,

d_j^+ : overachievement of targeted goals,

d_j^- : underachievement of targeted goals.

The priorities for the problem are as follows:

P_1 : to maximize the targeted profit per month,

P_2 : to satisfy the targeted demand per month,

P_3 : to minimize feeding cost per month,

P_4 : to minimize medical cost per month.

From the formulation above, equation (1) shows the objective function which is minimizing the deviations from the targeted goals. Equations (2) to (5) are the goal constraints while equations (6) to (7) are constraints related to the availability of resources. Equations (8) and (9), on the other hand, are the lower bounds and upper bounds respectively, whereas equation (10) denotes the non-negativity constraints. This model as mentioned is for the non-festive season model.

B. Goal Programming Model for Festive Season

The model for festive season can also be formulated. The definition of festive season here denotes the months where Muslims use to celebrate "hariraya puasa" or "hariraya qurban" in Malaysia. Normally, the demand of live goats is normally higher by 20 to 25%. For the year 2012, the festive months are August, September and October where both "hariraya puasa" and "hariraya qurban" fall within these months. It is therefore appropriate to utilize this model as different from the previous non-festive season model. This is due to the difference in demand for both seasons.

The same type of problem formulation is carried out for the festive season model as done for the non-festive season model. However, the values of the targeted goals as well as some of the right-hand side values of the resources availabilities are also higher.

C. Sources of Data

All the data for this study is collected from both the management office as well as the site of the farm. A week's stay in the farm and the mingling of the research team with the workers and supervisors of the farm gives an added advantage to get valuable information as well as the required data.

VI.   RESULTS AND DISCUSSIONS

A. General Optimum Results

Using SAS/OR software the goal programming models for both non-festive and festive seasons can be solved. The summary of the results can be depicted in Table I.

TABLE I: SUMMARY OF THE GOAL PROGRAMMING SOLUTIONS FOR NON-FESTIVE AND FESTIVE SEASONS

Variables	Entity	Non-Festive Season (Model 1)	Festive Season (Model2)
X_1	Boer Goat	40	70
X_2	Jamnapari Goat	60	30
X_3	Katjang Goat	120	120
X_4	Sheep	100	180
X_5	Cross Breed Boer-Jamn.	50	50
-	Profit	RM 120,000	RM 160,000

*Non-festive season: $P_1=0, P_2\neq0, P_3=P_4=0$, $d_2^-=30$

* Festive season: $P_1\neq0, P_2=P_3=P_4=0$ $d_1^-=12,800$.

As shown from Table 1, all values of Ps for the non-festive model are equal to "0" except P_2 is not equal to "0". This means that goals 1, 3 and 4 are achieved, while goal 2 is not achieved, which is shown by the value of $d_2^- = 30$. Thus, the non-festive seasons, the optimum supply of goats per month is recorded as 40 for Boer goat, 60 for Jamnapari goat, 120 for Katjang goat, 100 for sheep and 50 for cross-breed Jamnapari-Boer goat. The targeted profit of RM 120,000 per month is also achieved.

For the festive season model, on the other hand, the optimum supply per month is recorded as 70 for Boer goat, 30 for Jamnapari goat, 120 for Katjang goat, 180 for sheep and 50 for cross-breed Jamnapari-Boer goat. The results suggest that the supply of boer goat and sheep is almost double in the festive months. It makes sense, as boer goat and sheep are preferred to be slaughtered for "qurban" in the festive months. Nonetheless, the festive season model shows all goals are achieved with the exception of the profit-goal is underachieved by RM 12,800 as depicted from the results that $P_1\neq 0$ and d_1^- =12,800. However, the value of the under-achievement for this target is not high (about 8%) below target.

B. Constraint Analysis

The results can further be explained using part of the constraint analysis. It is done to determine whether there exists

some realignment of resources to reach the optimum solution given above. The analysis for the non-festive season is given in Table II.

Table II depicts that the targeted profit is over-achieved by RM 1,100 as d_1^+=1,100. However, demand is underachieved by 30 goats and sheep compared to the targeted demand of 400. This is shown by the value of d_2^- equals to 30. On the other hand, the feeding cost for goat is underachieved by RM 5,480. This can be seen from the results that the value of d_3^- is 5,480. This means that the total cost of feeding for goat is actually only RM 12,620 per month compared to the targeted feeding cost of RM 18,100.

The medical cost per month, is also underachieved as d_4^- is equal to 870. By looking at the utilization of the other resources, like total labor and overtime costs, it seems that they are underachieved.

TABLE II : CONSTRAINT ANALYSIS FOR NON-FESTIVE SEASON

Constraints/goals	RHS value	d_+ (row-i)	d_- (row-i)
G1/profit	120,000	1,100	0
G2/satisfying demand	400	0	30
G3/feeding cost	18,100	0	5,480
G4/medical cost	1,500	0	870
Labor cost	2,100	0	1,059
Overtime cost	945	0	375

For the festive season, on the other hand, part of the relevant constraint analysis is shown in Table III below:

TABLE III: CONSTRAINT ANALYSIS FOR FESTIVE SEASON

Constraints/goals	RHS value	d_+ (row-i)	d_- (row-i)
G1/profit	160,000	0	12,800
G2/satisfying demand	450	0	0
G3/feeding cost	21,100	0	6,320
G4/medical cost	1,800	0	944
Labor cost	2,300	0	1,035
Overtime cost	945	0	255

From Table III, it reveals that the profit goal of RM 160,000 for festive season is underachieved by RM 12,800 as d_1^- =12,800. The demand however, is directly achieved as both d_2^+ and d_2^- are equal to zeroes. In retrospect, feeding cost and medical cost for goats and sheep are underachieved by RM 6,320 and RM 944 respectively as shown by the values of d_3^- and d_4^-. It appears that the costs are lower than the expected or targeted costs.

In the case of labor and overtime cost for handling activities for goat, it shows that they are under-utilized as shown by the positive values of d_5^- and d_6^-.

VII.   CONCLUSION

The study reveals that the goal programming approach can be applied in solving this smallholder's farm management and operational problem. The results indicated that by applying both the non-festive and festive models, we can distinguish different strategies in fulfilling the demands at different seasons. As can be seen from the constraint analyses from previous Tables II and III, the number of live goats and sheep that can be supplied during the non-festive season is below the targeted demand but during the festive season, it could frulfill the demand. Thus, it makes sense to strategize by adopting aggressive marketing for promoting demand in both seasons. It is also sanguine to embark in enlarging the farm for increasing supply of goat and sheep, especially during the festive season.

The results also gave an opportunity for farm management to evaluate its position whether to maximize profit or minimize operational costs. However, looking at the scenario of the problem with conflicting goals, it is customary to adopt an act which satisfies all the goals. In other words, getting a compromising solution is a better choice than getting the maximum values of profit only.

The study also gives an opportunity for the research team to interact fully with the workers in the smallholder's farm where theoretical knowledge can be operationally applied in farm management. The awareness of making decisions when confronted with conflicting and multiple goals given the existing constraints of resources would be an added advantage to the management of farm. Last but not least, this exercise can be applied to many other similar smallholder's farm nation-wide.

REFERENCES

[1] M. Herreroa, R.H .Fawcetta, J.B Dent, "Bio-economic Evaluation of Dairy Farm Management Scenarios using Intergrated Simulations and Multiple Criteria Models," Elsevier Journal of Agricultural Systems, Vol 62, 1999, pp 169-188.

[2] I. Gribkoskaia, O.B Gullberg, K.J Horden, "Optimization Model for a Livestock Collection Plan," International Journal of Physical Distribution and Logistic Management, Vol 36, No 2, 2006, pp 136-152.

[3] B.W Taylor, P.F Anderson," Aspects of Goal Programming Approach to Marketing/ Production Planning ," Industrial Marketing Management, Vol 8, 1979, pp 136-144.

[4] Economic Planning Unit, Prime Minister's Department, Malaysia, "10th Malaysia Plan 2011-2015," 2010.

[5] Economic Planning Unit, Prime Minister's Department, Malaysia "9th Malaysia Plan 2006-2010," 2006.

[6] Ministry of Agriculture, Malaysia, "NAP3: New Agricultural Policy 3, 1998-2010," 1998.

[7] T.Rehman, P. Lara, C. Romero," Livestock Ration Formulation and Multiple-Criteria Decision making Techniques: A Review and Future Prospects," Proceedings of the 10th International Conference on Multiple-Criteria Decision Making, Vol 2, Taipei, Taiwan, 1992, pp 87-96.

[8] N.K. Kwak, M.J. Schneiderjans, " An Application of linear Goal programming to the Market Distribution," European Journal of Operational research," Vol 32 1993, pp 334-344.

[9] M.S. Sauian, "A Goal Programming Approach in Strategizing Production in a Plastic Manufacturing Firm," 7th. International Conference on Multi-Objective and Goal Programming, (MOPGP06), Tours, France, June 2006.

[10] B. Vitoriano, T. Ortuno, G.Tirado," HADS, a Goal Programming-Based Humanitarian Aid Distribution System," Journal of Multi-Criteria Decision Analysis, Vol 16 (2009), John Wiley & Sons, 2010, pp 55-64.

ISBN 978-952-265-435-9

9 789522 654359 >

Collan, Mikael; Hämäläinen, Jari; Luukka, Pasi (Eds.)
Proceedings of the FORS40 Workshop
LUT Scientific and Expertise Publications No. 13, 2013
Research Reports - Tutkimusraportit
ISSN-L 2243-3376
ISSN 2243-3376
ISBN 978-952-265-435-9 Printed publication

1. Operations Research

www.ingramcontent.com/pod-product-compliance
Lightning Source LLC
Chambersburg PA
CBHW051414200326
41520CB00023B/7235